AF869110

THE NEW
HEAT TRANSFER

SECOND EDITION

1989

EUGENE F. ADIUTORI

VENTUNO PRESS
6792 TIMBERWOOD DR.
WEST CHESTER, OHIO 45069

Copyright © 1974 by Eugene F. Adiutori.

Copyright © 1989 by Eugene F. Adiutori.

All Rights Reserved. Printed in the United States of America. No part of this publication may be reproduced, stored in a retrieval system, or transmitted, in any form or by any means, electronic, mechanical, photocopying, recording, or otherwise, without the prior written permission of the publisher.

for all who stand up for Freedom

PREFACE TO THE SECOND EDITION

This book describes a new engineering science based on the simple idea that:

> *Engineering sciences should be founded on free form descriptions of behavior, using fundamental parameters such as temperature, pressure, heat flux, flow rate, volts, stress.*
>
> *They should have NOTHING TO DO with contrived parameters such as electrical resistance, heat transfer coefficient, friction factor, elastic modulus.*

Today's engineering sciences are founded on "Laws" which focus our attention on contrived rather than fundamental parameters. For example, Ohm's Law and its "resistance" parameter contrived by dividing volts by amps; "Newton's Law of Cooling" and its "heat transfer coefficient" parameter contrived by dividing heat flux by boundary layer temperature difference.

The new engineering sciences abandon all "Laws" and all contrived parameters. The Laws are replaced by free form expressions which contain only fundamental parameters. The table below describes the basic difference between old and new engineering sciences:

OLD ENGINEERING FORM	NEW ENGINEERING TRANSFORMATION
$V = IR$	$V = f\{I\}$
$q = h\,\Delta T$	$q = f\{\Delta T\}$
$\sigma = E\,\epsilon$	$\sigma = f\{\epsilon\}$
R	V/I
h	$q/\Delta T$
E	σ/ϵ

The free form expressions of the new engineering little resemble the Laws of the old engineering, but they closely resemble the free form expressions which have been the foundation of pure mathematics and motion analysis for centuries:

$$y = f\{x\}$$

$$s = f\{t\}$$

The close resemblance to pure mathematics is important because it means that the new engineering does not require a new system of mathematics. It requires merely that the unsound mathematics of today's engineering be replaced by the mathematics of free form expressions and separated variables--the mathematics learned in pure mathematics.

In pure mathematics, the first step in the solution of equations is an attempt to "separate the variables". This step has a single purpose--to simplify the solution of the equation.

> In today's engineering sciences, problems are solved *without* separating the variables because contrived parameters such as R and h *combine* the variables which must be separated in order to simplify problem solution.

Electrical resistance is the ratio of volts to amps. In order to separate volts and amps, the contrived parameter called "resistance" must be *abandoned*.

The heat transfer coefficient is the ratio of heat flux to temperature difference. In order to separate heat flux and temperature difference, the contrived parameter called "heat transfer coefficient" must be *abandoned*.

> *The particular advantage of the new free form expressions is that they keep the important parameters separate, and this makes it possible to solve problems in the simplest way.*

The new electrical science has nothing to do with the parameter called "resistance", and therefore there is nothing to prevent the separation of volts and amps--nothing to prevent the solution of electrical problems in the simplest possible way.

The new heat transfer science has nothing to do with the parameter called "heat transfer coefficient", and therefore there is nothing to prevent the separation of heat flux and temperature difference--nothing to prevent the solution of heat transfer problems in the simplest possible way.

As in pure mathematics, if a problem deals only with proportional behavior, it makes little difference whether the parameters are separated or combined. But if the problem involves highly nonlinear behavior, the difference between separated and combined often makes the difference between a solution which is quite simple, and a solution which is virtually impossible.

When the engineering Laws and their contrived parameters were formulated many generations ago, they worked well enough because the world was a proportional place. There were no vacuum tubes--no semiconductors. Two phase heat transfer and fluid flow had little commercial significance. Plastic deformation had no practical importance.

But today, nonlinear behavior has immense practical importance, and the Laws and their contrived parameters do *NOT* work well enough--they greatly complicate the solution of many practical problems, and they make some so difficult that they have defied solution for *decades*.

In short, the old engineering and its Laws and contrived parameters must be abandoned because they complicate the solution of practical problems. They must be replaced by free form expressions and fundamental parameters in order to make it possible to separate variables and to solve engineering problems in the simplest way.

This book describes the new engineering sciences based on free form expressions and fundamental parameters. In particular, it:

- Presents a detailed description of the new heat transfer science, and demonstrates its application to the solution of practical problems. The new heat transfer science is based on the free form concept of "thermal behavior". This new concept replaces the old concepts of thermal conductivity, heat transfer coefficient, and radiative emissivity.

- Presents a detailed description of the new electrical science, and demonstrates its application to the solution of practical problems which involve purely resistive behavior. The new electrical science is based on the free form concept of "electrical behavior". This new concept replaces the old concepts of resistance, capacitance, and inductance.

- Presents free form expressions which are the bases for other branches of the new engineering.

- Serves as a pilot for books on other branches of the new engineering.

This book is a second edition, but it is more a sequel than a second edition. It is a complete rewrite of the first edition written fifteen years ago. I have rewritten it with one goal in mind: To make it better. And since it could be better only if it was different, it is different. The science is the same, but the explanation of the science is altogether different. The first edition is more a description than a demonstration. This edition is more a demonstration than a description.

This book is designed for self-instruction. It contains a large number of chapters or modules which are more or less independent and, with a few exceptions, can be completed in one sitting. The book is in five parts. It is intended to be read from cover to cover, and then to be used for reference. The table of contents will serve as a subject index.

As in the first edition, the language and the mathematics are as simple and straightforward as I could make them. I vigorously attack current methods in order to promote methods which are new and better. For this, I make no apology. If anyone is personally offended by my attacks on current methods, the offense is not intended and should be overlooked. My concern is with science and methods--not with personalities and politics.

I am well aware that my proposal to replace today's engineering sciences with sciences I have conceived is quite preposterous. But that proves nothing. If preposterous ideas were not an integral part of science, the earth would still be flat. Perhaps my proposal will seem less preposterous after you have read the first chapter which describes a few of the "impossible" problems I have solved with the new engineering.

In summary, the new engineering sciences presented here are better because they are much simpler than today's engineering sciences. And because science is the search for a better way, I have no doubt that these new engineering sciences will be everyday engineering sciences tomorrow.

8/22/89

PREFACE TO THE FIRST EDITION

This book presents a new concept of heat transfer--a new way to think about heat transfer, a new way to correlate heat transfer experimental results, a new way to design heat transfer equipment, a new way to understand the behavior of heat transfer equipment. This new way will replace Fourier's way because the new way is simpler, more logical, and much more powerful. Any problem which can be solved with Fourier's concept of heat transfer can be solved more simply, more reliably, and more accurately with the new concept. And problems which have defied solution with Fourier's heat transfer are solved with dispatch using the new heat transfer.

I have written this book for engineers and educators--and anyone interested in science. It is neither a textbook nor a handbook. It is not intended to impress the reader with my erudition or to dumbfound him with mathematics. It is an attempt to describe the new heat transfer and its application to engineers and educators who are familiar with the old heat transfer. And I have tried to present the new heat transfer in such a way that educators could teach it and engineers could apply it at the same time the leaders of the scientific community are debating its pros and cons.

I well recognize (and have frequently been told) that my writing style little resembles twentieth century scientific prose. It is the style I prefer--and the style which to me seems best suited to my goal. I wish to be understood--and a clear understanding demands clear, straightforward language.

I know that many will imagine themselves offended by my critical examination of theories and concepts of long standing--particularly when this examination demonstrates that the old ways are no longer useful and must be abandoned. I regret the offense--it is no part of my purpose.

Many times in this book, I have had to choose between possible offense on the one hand and science on the other. Each time, I have chosen science. Any error in this book is an honest error--I have not propagated a single myth, correlation, concept, or conclusion for expediency. There is nothing in this book which I do not firmly believe--nor have I expressed an opinion in any area where I do not feel well qualified to do so.

Much of this book is at odds with what has been considered accomplished scientific fact for many decades. I do not pass over the differences lightly--each time, I attempt to show how and why the new way is better than the old way. Sometimes it may seem I am mocking the old ways. I am not. My purpose is to dispel the old ways at the same time I present the new ways. In science, there has always been room for only one way--the best way at the time.

In short, this is a book about a new and simple science which will replace an old and frightfully complex art--and I have written it for engineers and educators that they may independently decide between Fourier's way and my way.

TABLE OF CONTENTS

PART 1 THE NEW HEAT TRANSFER--FUNDAMENTALS

PART 2 THE NEW ELECTRICAL SCIENCE

PART 3 THE GENERATION OF ENGINEERING SCIENCES

PART 4 THE NEW HEAT TRANSFER--EXPERIMENTATION

PART 5 THE NEW HEAT TRANSFER--NONLINEAR PHENOMENA

CHAPTER 1

HOW TO THINK IN THE NEW ENGINEERING SCIENCES

In the new engineering sciences, we think in a free form way, using only parameters which are fundamental. We *abandon* all Laws which rigidly prescribe behavior, and all parameters which are not fundamental--parameters created by combining other parameters--contrived parameters.

Equation (1-1) is an example of a free form expression. It identifies the fundamental parameters y and x, and makes no statement about functionality.

$$y = f\{x\} \qquad (1\text{-}1)$$

Today's old engineering has no use for free form expressions, and this equation would be rewritten in the form of a Law such as Eq. (1-2).

$$y = Hx \qquad (1\text{-}2)$$

Converting Eq. (1-1) to a Law has no effect on y and x, but it does create a new parameter--the parameter "H". The Law appears to state that y is proportional to x, and that H is the constant of proportionality. However, the appearance is deceiving.

In words reminiscent of today's old heat transfer, "This Law is not a phenomenological description of behavior. It is the defining equation for H." Meaning that y is NOT generally proportional to x, and H is NOT a constant in the equation. H is a *variable* in the equation--H is a contrived variable, created by dividing y by x.

It is important to note that Eqs. (1-1) and (1-2) are *exactly the same*. Both state that y is an unspecified function of x. The seeming difference results because Eq. (1-2) appears to make a statement about functionality, but does not actually do so.

It is also important to note that Eq. (1-1) contains only two variables--y and x. On the other hand, Eq. (1-2) contains *three* variables--y and x, and the extraneous variable H invented/created/contrived by dividing y by x.

In the new electrical science, we abandon Ohm's Law and the idea of electrical resistance--the idea that the ratio of volts to amps can be viewed as a "resistance" to the flow of electricity. We think of volts and amps individually--separately. And we replace Ohm's Law with the free form expression

$$V = f\{I\} \tag{1-3}$$

which states that V is an unspecified function of I. This function can and does take many different forms including proportional and nonlinear.

In the new heat transfer science, we abandon "Newton's Law of Cooling" and the idea of the heat transfer coefficient--the idea that the ratio of heat flux to temperature difference can be viewed as the inverse of resistance to the flow of heat. We think of heat flux and temperature difference individually--separately. We replace "Newton's Law of Cooling" with the free form expression

$$q = f\{\Delta T\} \tag{1-4}$$

which states that q is an unspecified function of ΔT. This function can and does take many different forms, including proportional and nonlinear.

In the new heat transfer science, we abandon Fourier's Law of Conductive Heat Transfer and the idea of thermal conductivity--the idea that the ratio of heat flux to temperature gradient can be viewed as the inverse of resistance to the flow of heat. We think of heat flux and temperature gradient individually--separately. We replace Fourier's Law of Conductive Heat Transfer with the free form expression

$$q = f\{dT/dx\} \quad (1\text{-}5)$$

which states that q is an unspecified function of dT/dx. This function can take many different forms, but today's view is that it is a proportional function for all known materials. Eq. (1-5) is a free form expression used to describe a phenomenon which universally behaves in a proportional way--*today*.

In the new material science, we abandon Young's Law and the idea of elastic modulus--the idea that the ratio of stress to strain in the elastic region can usefully be viewed as a material property. We think of stress and strain individually--separately. We replace Young's Law and all similar Laws and moduli with

$$\sigma = f\{\epsilon\} \quad (1\text{-}6)$$

In short, we abandon all behavioral Laws and all contrived parameters. We replace behavioral Laws with free form expressions which identify the parameters of importance, but make no statement about functionality. And we replace the contrived parameters in these Laws with the fundamental parameters from which they are contrived.

The end result is a new engineering science which is simple and clear. Simple because it contains only ideas and parameters which are necessary. Clear because it allows us to think and to solve problems with the important parameters *separated*.

PROVING THAT THE NEW WAY IS A BETTER WAY TO THINK

If the new engineering is a better way to think (*and it is*), then persons who use it should be better able to solve practical engineering problems (*and they are*). But how does one demonstrate that a new way of thinking is better than today's "old" way?

One way would be to show that someone who is familiar with both ways is readily able to solve practical engineering problems which defy solution with the old engineering. That is the method employed in this chapter. It describes several engineering problems I solved using the new engineering.

All of these problems were quite simple for me, as shown by the short length of time I required to solve them. These same problems are extremely difficult in the old engineering, as shown by the fact that several of them defied solution *for decades*.

Of course, I cannot *prove* that my thinking was based on new rather than old engineering. But I hope the reader will respond to the problems along the following lines:

- These problems seem difficult. If Adiutori solved them with little effort as he says, there must be something better about the way he solves problems. He claims this "something" is the fact that he thinks of free form expressions and fundamental parameters, and NEVER thinks of Laws and contrived parameters.

- Saunders' (1976) review in the International Journal of Heat and Mass Transfer indicates there may be something to Adiutori's claim: "It is to be hoped that many will read these volumes carefully . . . and that their influence will be brought to bear on the whole community of heat transfer workers . . ."

- The publication of the Russian edition of *The New Heat Transfer* by Mir (Moscow), the publishing house of the Russian government, gives credibility to Adiutori's claim.

- Since these problems suggest that there is much to be gained from the new engineering, I will read on in order to independently decide whether it is something I will use.

HEAT FLUX DISTRIBUTION IN FOSSIL FUEL POWER PLANTS

In the early 1970's, I was a consultant to a company which manufactures components for fossil fuel power plants. In the course of my work, I was asked to review the design/manufacture/installation/operation of their fossil fuel plant components in the hope of finding something useful.

During a field trip to an operating power plant, I observed several field service reps inserting a 20 ft. thermocouple probe into the boiler. Data were being obtained to compare with design data, and to generate an improved data bank. Since the boiler was greater than 100 ft. wide, the temperature distribution could be measured in less than half of the boiler volume. (The probe used was water-cooled and manually held, and a probe length greater than 20 ft. would not be practical.)

The temperature distribution in the boiler is important because, together with view factor and emissivity, it determines the heat flux at the surface of the boiler components. However, it would be much more desirable to measure the heat flux *directly* because this would reveal the combined effects of temperature distribution, view factor, and emissivity.

The above line of reasoning suggested the following problem statement:

> Is there a *simple* way to measure the heat flux distribution in a fossil fuel boiler across the full 100+ ft. width?

Using the new engineering to think about this problem, I invented a very simple way to measure the heat flux distribution across the full width of fossil fuel boilers. The method is accurate and reliable, and is extremely simple.

There is something about my method which you may find quite difficult to believe:

> The method does NOT involve inserting sensors into the boiler! The method utilizes only hardware which is an integral part of large, fossil fuel boilers.

My method does not utilize optical pyrometers or any high tech equipment. It utilizes only the boiler and its components, and very simple equipment and instruments found in any large fossil fuel power plant. Nothing is modified inside the boiler, and only very simple things are done outside the boiler *while it is operating.*

I had only a few months experience with fossil fuel power plants when I invented the subject method. For me, the invention of the method was quite simple. But only because I used the new engineering to think about the problem.

If you are interested in the method, please refer to U. S. Patent No. 3,901,081, dated August 26, 1975, inventor Eugene F. Adiutori.

In order to demonstrate the accuracy of this new method, it was necessary to invent a completely independent method of measuring heat flux. This also was a simple task--using the new engineering.

This other method is based on converting the above temperature probe into a heat flux probe. (The other method has not been patented, perhaps because it is so simple.)

The heat flux data obtained by the two completely independent methods in the 20 ft. range of the modified probe showed excellent agreement, and demonstrated the accuracy of both new methods.

NEW TYPE OF HEAT FLUX GAGE

A few years ago, I was assigned an experimental program which needed a type of heat flux gage which did not exist. The required gage had to have the following attributes:

- Small size--less than .025" x .025".
- Accurate.
- Able to withstand corrosive, erosive environment to 2000 F.
- Rugged--able to withstand high g loads.
- Simple to make.
- Simple to use.
- Inexpensive.

I had used the new engineering to think about this problem for less than a week when I invented a new type of heat flux gage which has *ALL* the above features. We fabricated and calibrated 50 of the gages in a poorly equipped heat transfer laboratory. The gages made it possible to accurately measure the heat flux distribution in the test hardware with a resolution and accuracy far better than had ever before been attained in similar test programs reported in the literature.

This new type of heat flux gage should become a universal standard in the heat transfer world. It is currently in the patent process of a large corporation, and therefore I am not at liberty to describe it. If you are interested in the gage, please monitor the publications of the U. S. Patent Office for inventions by Eugene F. Adiutori.

NEW TYPE OF THERMAL CONDUCTIVITY METER

Several years ago, a need arose to measure the as-built "thermal conductivity" of a relatively thin film of low conductivity material. Since we did not have an apparatus to measure thermal conductivity, we contacted a university which had a research institute engaged in the business of measuring thermal conductivity. The research institute had the standard, commercial apparatus which costs $30,000.

When we described what was needed, the institute personnel replied that they were unable to measure the thermal conductivity of thin films. They could measure the thermal conductivity of thick specimens, but that was not what was needed.

I had used the new engineering to think about this problem for less than a week when I invented a new type of thermal conductivity meter. The meter was built in a heat transfer laboratory in one day at a cost of less than $1,000. And it was used to measure the thermal conductivity of numerous thin and thick films with very good accuracy.

This new type of thermal conductivity meter is in the patent process of a large corporation, and therefore I am not at liberty to describe it. If you are interested in the meter, please monitor the publications of the U. S. Patent Office for inventions by Eugene F. Adiutori.

PNEUMATIC TRANSPORT SYSTEM

Several years ago, I was a consultant to a company which manufactures a variety of products including pneumatic transport systems used to convey metal pellets. I was asked to review the design/installation/operation of the pneumatic transport system to determine whether/how the system capacity could be increased from the then current value of 150 #/min.

The system design was quite conventional and consisted of a large compressor, a venturi which induced the pellets into the transport line, and open loop piping.

In order to determine whether it would be possible to increase the system capacity, I used the new engineering to design a very simple field test which would provide the answer in a definitive way. The test data indicated that a major increase in system capacity would be possible using the existing compressor and piping.

I then used the new engineering to devise a very simple way to increase the system capacity. The modification involved replacing the venturi with the component I designed, and no change was made in the compressor or the piping. A field installation was modified to the new configuration, and the very first field test indicated that the system capacity was greater than 500 #/min!

That one experience was the only experience I have ever had with pneumatic transport systems. The entire experience covered a time span of approximately four weeks. Using the new engineering, I was able to solve this system design problem in four weeks, when 20 or 30 years had not been long enough using today's old engineering.

DRAG COMPONENTS IN A NUCLEAR CONTROL ROD DRIVE

In the early 1970's, I was responsible for the motion analysis of the control rod drive for a pressurized water, nuclear power plant. The design specifications for the drive required a minimum insertion rate (ie control rod movement vs time) following a scram (ie a rapid reactor shutdown in which the control rod free falls into the reactor core).

To meet the insertion rate specifications, it was necessary to ensure that the drag forces on the control rod were not excessive during the free fall. To accomplish this, design limits were established for dynamic mechanical drag and hydraulic drag. Since different company departments were responsible for the two types of drag, the following problem arose:

> When the control rod drive has been built and the insertion rate following a scram has been measured, what will we do if the measured insertion rate does not satisfy the design specification?

> How will we know whether the problem is dynamic mechanical drag or hydraulic drag? How will we know which company department owns the problem?

I devised a very simple solution to this problem. The solution involves measuring both the dynamic mechanical drag and the hydraulic drag by analyzing the distance vs time data obtained in measuring the as-built insertion rate!

The method is simple and accurate, and is described in the literature. If you are interested in the method, please see *Nuclear Engineering and Design* 30 (1974) 144.

PUBLICLY UNDOCUMENTED RESULTS

I have used the new engineering since 1963. Like the pneumatic transport problem described above, most of the problems I have solved with it are described in company files not accessible to the public. The following accurate account is not documented in the open literature, but it is well documented in the files of a large corporation and in my personal file.

I was asked to work on an unsolved problem which had consumed all of the following: several man months of effort by engineers highly competent in the old heat transfer; many hours of main frame computer time; a $25,000 experimental program. The essence of the problem was:

> Why is the as-built "heat transfer coefficient" 400% greater than our analysis predicts?

(Since I live and work in a world based on the old heat transfer, I necessarily communicate in the language of the old heat transfer. But although I speak and write both new and old languages, my thinking is exclusively new way.)

I used the new engineering to think about this problem--to deduce the phenomenon which caused the discrepancy between analytical result and reality. I solved it *in less than ten minutes*! And my solution was obviously correct because it was so simple.

THERMAL CHECK

A common method of cooling or heating equipment is to provide internal cooling/heating circuits through which fluid is circulated. If the thermal performance of the cooling/heating circuit is critically important, it would be highly desirable to measure the as-built thermal performance of the circuit in a component test.

Such a measurement would be extremely valuable because it would make it possible to determine whether the as-built hardware satisfied the intended thermal performance of a new circuit design *as soon as the first part was made*.

The normal method of accomplishing this is to test the component as part of a prototype test of the entire system. The difficulty with the normal method is that one or two years is oftentimes required to build and test a prototype system, in which case the test results would not be obtained until one or two years *after* the first part was made. This delay directly impacts on the design cycle, and the elimination of this delay represents a major improvement in the design process.

A component measurement also would provide an inexpensive method for generating a data bank, and a useful tool for quality control.

I had used the new engineering to think about this problem for a short time when I invented a method of measuring the as-built thermal performance of cooling/heating circuits in a component test. I chose to call the method "thermal check" because it is the thermal analog of "flow check", a standard method of measuring as-built fluid flow performance in a component test.

Thermal check should become an industry standard because it is simple, accurate, and inexpensive. It is currently in the patent process of a large corporation, and therefore I am not at liberty to describe it. If you are interested in thermal check, please monitor the publications of the U. S. Patent Office for inventions by Eugene F. Adiutori.

THERMAL STABILITY

In 1963, the stability of the heat transfer process was not within the knowledge envelope of heat transfer science. At that time, I was in charge of a forced convection, boiling liquid metal test facility for a six month period. The facility had several baffling stability problems which defied solution using the old heat transfer.

During that six month period, I abandoned heat transfer coefficients. And after I had taken that first step toward the new engineering, it was a matter of only a few weeks until I had solved the problem of heat transfer stability in a general way and also in several particular ways.

I chose to give the generic name "thermal stability" to the stability of the heat transfer process. Adiutori (1964a) and (1964c) briefly describe my solution of the thermal stability problem and point out several practical and important applications of that solution.

DESIGN PATENTS

I have used the new engineering to design the thermal and fluid aspects of a variety of hardware, and several new design concepts have resulted in patents. The design patent in which I take particular pride is U. S. Patent No. 3,388,888 dated June 18, 1968. It concerns a component which promotes cooling, and is now found in virtually all aircraft gas turbines.

THINKING

Inventions and methods and problem solutions all result from *thinking*. The results of my thinking are related in this chapter for only one reason:

> To suggest to the reader that I must *think* in a different and better way in order to have solved the above problems quickly/easily/simply.

And I *DO* think in a different and better way. I think more clearly because I think with far fewer parameters. I am better able to cope with real things because I think *ONLY* of real things. I allow *NO* contrived parameter to invade/dilute/pollute my thinking.

Is it easy to learn my way of thinking? Yes! You merely have to learn how to deal with engineering problems using pure mathematics and fundamental parameters. And you have to forget the ineffective mathematics and the Laws and the contrived parameters which are the bases of today's old engineering.

You are already familiar with the pure mathematics and the fundamental parameters which are the bases of the new engineering. You merely have to learn how to apply the mathematics you already know, to the solution of problems with which you are already familiar, using parameters you now use every day.

In a very real sense, you already know and understand all the components of the new engineering. You merely have to learn how to put them together in the new way described and demonstrated in this book.

CHAPTER 2

A BRIEF SUMMARY OF THE NEW HEAT TRANSFER

The new heat transfer is based on the new, free form concept of "thermal behavior". The particular advantage of this new science is that it deals effectively with all forms of heat transfer behavior including highly nonlinear.

Today's old heat transfer is based on the concepts of thermal conductivity (k) and heat transfer coefficient (h). The particular disadvantage of the old heat transfer is that it deals effectively only with behavior which is strictly proportional.

WHY THE NEW HEAT TRANSFER?

The new heat transfer is a better way to think/correlate/design/analyze. It is better because it is simpler. And because it is simpler, it improves the solution of everyday problems, and makes possible the solution of certain practical problems never solved with the old heat transfer.

Today's old heat transfer is a modified form of the heat transfer science conceived by Fourier (1822). His experiments indicated that heat transfer phenomena generally behave in a proportional way, and therefore he conceived a science which effectively deals with proportional behavior.

> *Neither Fourier's original science nor today's modified version deals effectively with nonlinear behavior.*

In the twentieth century, nonlinear heat transfer phenomena have become increasingly important, a trend which will surely continue into the twenty-first century. Because the new heat transfer deals effectively with nonlinear behavior whereas today's old heat transfer does not, the latter will soon be abandoned.

HOW IS THE NEW HEAT TRANSFER SIMPLER?

In both concept and application, the new heat transfer is much simpler than the old heat transfer.

The conceptual simplicity of the new heat transfer results from the elimination of the many thermal parameters which are unnecessary. For example, the old parameter called "heat transfer coefficient" is obviously unnecessary because it is contrived--it is a shorthand way of writing "the ratio of heat flux (q) to boundary layer temperature difference"--a shorthand way of writing $q/\Delta T$.

In the new heat transfer, we do not use $q/\Delta T$ (alias h). We use $q\{\Delta T\}$--thermal behavior--the functional relationship between q and ΔT. Similarly, thermal conductivity and thermal emissivity are eliminated and are replaced by thermal behavior expressions in the forms $q\{dT/dx\}$ and $q\{T\}$. Notice that this eliminates three of the old parameters, and that they are not replaced by new parameters. They are replaced by old parameters used in a new way.

The conceptual simplicity of the new heat transfer also results from the elimination of all group parameters such as Nu, Re, f, St, etc. In the new heat transfer, we have groups of parameters, but we have no group parameters. For example, we might have xG/μ, but we would not have Re.

In application, the new heat transfer is simpler because problems are solved with the heat flux (q) and the thermal driving force (TDF) *separate*. In the old heat transfer, problems are solved without separating q and TDF because parameters like h and k are disguised forms of the ratio of q to TDF. h is $q/\Delta T$, and k is $q/(dT/dx)$. The end result is that problems involving nonlinear heat transfer are *much* simpler to solve using the new heat transfer.

WHAT IS THERMAL BEHAVIOR?

Thermal behavior has nothing to do with h, or k, or ϵ. These old parameters play no role in the new heat transfer. They are relegated to history books where they take their rightful places alongside phlogiston and caloric fluid.

In the new heat transfer, heat transfer is viewed as a dynamic process in which heat flux responds to thermal driving force (TDF). In thermal convection, the TDF is the temperature difference (ΔT) across the boundary layer. In thermal conduction, the TDF is the temperature gradient (dT/dx) in the material. In thermal radiation, the TDF is the surface temperature of the radiating body.

"Thermal behavior" is the functional relationship between heat flux and thermal driving force. Thermal behavior is $q\{\Delta T\}$ in convection, $q\{dT/dx\}$ in conduction, and $q\{T\}$ in radiation. (The symbolism "$q\{\Delta T\}$" states "q is a function of ΔT".)

In the new heat transfer, we deal with q and TDF in the same way we deal with y and x in mathematics--the same way we deal with distance (s) and time (t) in motion analysis. The basis of all mathematical analysis is

$$y = f\{x\} \tag{2-1}$$

The basis of all motion analysis is

$$s = f\{t\} \tag{2-2}$$

In the new heat transfer, the basis of all thermal analysis is:

$$q = f\{TDF\} \tag{2-3}$$

Equation (2-3) is the generic expression for *all* forms of thermal behavior. It is the basis for all thinking, experimenting, correlating, designing, and analyzing in the new heat transfer.

ABANDONING "DIMENSIONAL CONSISTENCY"

From the viewpoint of the old heat transfer, there is a serious difficulty with Eq. (2-3). The difficulty is that this equation is not "dimensionally consistent"--the dimensions on the left side of the equation do *not* equal the dimensions on the right side. Therefore Eq. (2-3) would be considered flawed and unscientific in the old heat transfer.

But from the viewpoint of the new heat transfer, Eq. (2-3) is sound and scientific. We have no use for dimensional consistency because Nature is not dimensionally consistent. The requirement that fundamental equations must be "dimensionally consistent" is an intellectual requirement which seems necessary because of the way we define "equal" and *not* because it is indicated by Natural phenomena.

There is nothing dimensionally consistent about the fact that a temperature difference causes a heat flux--that a voltage difference causes an electrical current--that a pressure difference causes a fluid flow. Eq. (2-2) is not dimensionally consistent, yet it is widely regarded as a sound and scientific description of Natural behavior.

The fundamental equations of the old heat transfer are dimensionally consistent *by design*. For example, Fourier intended the equation "$q = h\ \Delta T$" to be an expression of the fact that his experiments indicated q was proportional to ΔT. Fourier felt that "equal" meant "numerically equal and dimensionally identical", and so he felt that his equation should be dimensionally consistent.

Fourier achieved dimensional identity by *arbitrarily assigning* dimensions to the constant of proportionality between q and ΔT. That is how dimensional consistency is generally achieved--*by arbitrarily assigning dimensions to what are in fact pure numbers*.

In the new heat transfer, we deem it a useless exercise to assign dimensions to pure numbers, and so we abandon the intellectual requirement of "dimensional consistency". We use dimensions, but we are not concerned with dimensional consistency. More often than not, the equations we use are *NOT* dimensionally consistent--just as Nature is not dimensionally consistent--just as Eq. (2-2) is not dimensionally consistent.

THE NEW MEANING OF "EQUAL" AND THE EQUAL SIGN

In today's old heat transfer, "equal" has the same meaning it had for Fourier--*numerically equal and dimensionally identical*. Equations are contrived to be dimensionally identical by arbitrarily assigning dimensions to what are in fact pure numbers.

In the new heat transfer, we recognize that Natural phenomena are not dimensionally consistent, and so we abandon the unnatural requirement of dimensional identity. We define "equal" in a new way:

> *The word "equal" and the symbol "=" mean numerically equal. They make NO statement about dimensional identity.*

In the new heat transfer, parameter symbols identify the parameter *and its dimensions*. Since the symbols indicate dimensions, it would be redundant to express dimensions in the manner of the old heat transfer--ie it would be redundant to write q = 145 Btu/hr ft^2. We write "q = 145", and the nomenclature defines the symbol "q" to mean heat flux *in Btu/hr* ft^2.

For example, in the new heat transfer, the equation

$$q = 34.5\ \Delta T^{1.67}$$

means "q in its specified dimensions is numerically equal to 34.5 times ΔT in its specified dimensions raised to the power of 1.67". Because the new equations make no statement about dimensional identity, they cannot be dimensionally incorrect per se.

FUNDAMENTALS IN THE NEW HEAT TRANSFER

In the new heat transfer, Eq. (2-3) is the generic form for all types of heat transfer. Eqs. (2-4) to (2-6) are the specific forms for convection, conduction, and radiation:

$$q = f\{TDF\} \tag{2-3}$$

$$q_{conv} = f\{\Delta T\} \tag{2-4}$$

$$q_{cond} = f\{dT/dx\} \tag{2-5}$$

$$q_{radn} = f\{T\} \tag{2-6}$$

Note that these fundamental expressions of the new heat transfer contain only *two* thermal concepts:

- heat flux
- temperature

Heat flow systems are analyzed using Equations (2-4) to (2-6) and Equation (2-7):

$$\Delta T_{total} = \sum \Delta T_i \tag{2-7}$$

Equation (2-7) is obviously true and requires no structured proof. It states that the total temperature difference between any two points is the sum of the individual temperature differences between those points.

For example, the total temperature difference between two fluids separated by a wall equals the sum of three temperature differences: the temperature difference across the wall plus the temperature difference across each of the two fluid/solid interfaces. In this case, Eq. (2-7) takes on the specific form of Eq. (2-8):

$$\Delta T_{total} = \Delta T_1 + \Delta T_w + \Delta T_2 \qquad (2\text{-}8)$$

Equation (2-7) is of cardinal importance in the new heat transfer. It is the *keystone* which makes it possible to determine overall thermal behavior from a description of component thermal behavior. It is the replacement for old heat transfer expressions like

$$U = \frac{1}{1/h_1 + 1/h_2 + t/k} \qquad (2\text{-}9)$$

It is important to note that Eqs. (2-8) and (2-9) are *IDENTICAL*. The only difference is language. Eq. (2-8) is written in the language of the new heat transfer--the clear and simple language of separated variables. Eq. (2-9) is written in the language of the old heat transfer--the foggy and bewildering language of combined variables.

FUNDAMENTALS IN TODAY'S OLD HEAT TRANSFER

The fundamental expressions in today's old heat transfer are:

$$q_{conv} = h \, \Delta T \qquad (2\text{-}10)$$

$$q_{cond} = -k \, dT/dx \qquad (2\text{-}11)$$

$$q_{radn} = \epsilon \sigma T^4 \qquad (2\text{-}12)$$

These expressions contain *five* thermal concepts:

- heat flux
- temperature
- heat transfer coefficient
- thermal conductivity
- radiative emissivity

Eqs. (2-10) to (2-12) and equations similar to Eq. (2-9) are used to analyze heat transfer systems. Equation (2-9) makes it possible to determine the overall heat transfer coefficient from the individual coefficients, and therefore it is of cardinal importance in the old heat transfer.

There is no place for Equations (2-9) to (2-12) in the new heat transfer. They are altogether replaced by Eqs. (2-4) to (2-7).

A NEW/OLD DICTIONARY OF FUNDAMENTALS

The table below is a new/old dictionary of fundamentals which differ in the two heat transfers:

OLD HEAT TRANSFER	NEW HEAT TRANSFER
h	no analogous parameter
k	no analogous parameter
ϵ	no analogous parameter
$q = h\ \Delta T$	$q = f\{\Delta T\}$
$q = -k\ dT/dx$	$q = f\{dT/dx\}$
$q = \epsilon\sigma T^4$	$q = f\{T\}$
$U = \dfrac{1}{1/h_1 + 1/h_2 + t/k}$	$\Delta T_{total} = \Delta T_1 + \Delta T_2 + \Delta T_{wall}$

It is important to note that:

- *Every* problem that can be solved the old way with h, k, ϵ can be solved the new way with*out* h, k, ϵ.
- *Every* problem that can be solved the old way with Eqs. (2-9) to (2-12) can be solved the new way with Eqs. (2-4) to (2-7).

If the problem involves only proportional heat transfer behavior--ie if each thermal component is such that q is essentially proportional to ΔT (as in one phase, forced convection)--then the solution of the problem is equally simple using either the old or the new heat transfer.

But if the problem involves nonlinear heat transfer--if one of the thermal components exhibits even a moderately nonlinear relationship between q and ΔT (as in free convection)--then the solution of the problem is much simpler using the new heat transfer.

And if the nonlinearity is pronounced--if the q{ΔT} functionality of one of the thermal components is so highly nonlinear that it contains maxima and/or minima (as in boiling), then the difference in simplicity can and does make the difference between a solution which is essentially impossible the old way and quite simple the new way.

DIMENSIONAL ANALYSIS AND PARAMETERS OF THE INTELLECT

Dimensional analysis depends on the old heat transfer view that Natural phenomena are dimensionally consistent. But in the new heat transfer, we observe that Nature is *NOT* dimensionally consistent. Therefore there is no reason to suppose that the analysis of dimensions can reveal anything about parametric functionality.

In the new heat transfer, we have no faith in dimensional analysis, and results based solely on it are viewed with healthy skepticism. We consider that dimensional analysis is art--not science.

In the new heat transfer, we have no use for contrived parameters--parameters generated by combining groups of fundamental parameters--parameters not found in Nature, but found only in the intellect. We recognize that the invention of dimensional or dimensionless group parameters is harmful because the invented parameters replace rather than supplement the Natural parameters which are our real concern.

For example, the literature on the old heat transfer is replete with articles which "describe" experimental results only in terms of contrived parameters such as the "Nusselt number" and the "Reynolds number" From this dimensionless description, it is identically *impossible* to determine what was actually done or what was actually found in the experiment. And because our attention is focused on the functionality among contrived parameters, we lose sight of the functionality among the parameters which truly concern us--the fundamental parameters of the real world.

POWER LAWS AND LOG LOG GRAPH PAPER

The old heat transfer is based on the implicit assumption that Nature has a strong preference for power laws. This unfounded assumption is the reason power laws describe virtually all phenomena in the old heat transfer. As the later chapters demonstrate, power laws are oftentimes used *even though the data bear no resemblance to power laws*.

Since power laws are straight lines on logarithmic coordinates, the graph paper of choice is log log graph paper. That is why virtually all graphs in the old heat transfer literature and texts are log log graphs, even though it is essentially impossible to "see" the functionality suggested by data plotted on log log graphs.

In the new heat transfer, we recognize that there is no good reason to suppose that Nature prefers power laws--no good reason why log log paper should be the graph paper of choice. We generally plot data on *linear* graph paper so that we can readily "see" the functionality suggested by the data. And we correlate with whatever functionality is suggested by the data, including power laws.

In the old heat transfer, the use of log log graph paper can and does lead to experimental conclusions which little resemble the supporting data. The thermal behavior of nucleate boiling interfaces is a good example. For 40 years, it has been widely agreed that, in nucleate boiling, q is related to ΔT^n where n is "3 to 4". Yet the nucleate boiling data in the literature generally indicate a highly *linear* relationship between q and ΔT.

THERMAL STABILITY

"Thermal stability" is the generic name given to the stability of the heat transfer process by Adiutori (1964a). The subject of thermal stability is not within the knowledge envelope of the old heat transfer. The old heat transfer implicitly views heat transfer as a static process, and static processes do not have stability problems. The following by Rohsenow (1963) is indicative of the static view generally held in the old heat transfer:

> *With condensing vapor as the heat source on one side of a wall, any point on the entire (pool boiling) curve can be reached under stable conditions.*

This view was universal until the publication of Adiutori (1964a). Since then, others have recognized that this view is incorrect, and have published experimental evidence to that effect. However, the above view expressed by Rohsenow is still the view presented in the latest texts on heat transfer--for example, Kreith and Bohn (1986).

In the new heat transfer, heat transfer is viewed as a dynamic process--as a flow responding to a driving force. Since it is a dynamic process, it may occur in a stable or unstable manner. The generic criterion for thermal stability was first derived and presented by Adiutori (1964a).

As we shall see in the chapter on thermal stability, the above statement by Rohsenow should be corrected to read:

> With condensing vapor as the heat source on one side of a wall, a pool boiler may or may not be stable at all points of the pool boiling curve. If the boiler is unstable in a region of the boiling curve, the result will be hysteresis or undamped oscillatory behavior in that region.

The thermal stability of boilers depends on the thermal behavior of the boiling interface and also on the detailed design of the hardware. This is an important point because it indicates that the designer can exert a strong influence on the thermal stability of boilers in particular and heat transfer equipment in general.

HOW TO CONVERT HEAT TRANSFER COEFFICIENTS TO THERMAL BEHAVIOR CORRELATIONS

The new heat transfer involves abandoning heat transfer coefficients, but this does not mean that all the old heat transfer experiments must be repeated. It means only that h correlations must be converted to the $q\{\Delta T\}$ form of the new heat transfer. This is readily accomplished in the following way:

- Assume that the correlations of the old heat transfer accurately describe heat transfer behavior.

- Convert h correlations to thermal behavior correlations by eliminating all group parameters. They are readily eliminated by substituting the fundamental parameters from which they are contrived. For example, substitute xG/μ for Re, substitute $q/\Delta T$ for h.

- Separate q and ΔT--ie put q on one side of the equation, ΔT on the other side.

For example, the following h correlation is widely used to describe one phase, forced convection heat transfer:

$$Nu = .023\, Re^{.8}\, Pr^{.4} \qquad (2\text{-}13)$$

(Notice that this correlation is so opaque that it readily reveals nothing about anything of designer importance.) We convert this old way correlation by eliminating dimensionless group parameters like Nu and dimensional group parameters like h. The end result is

$$q = .023(G)^{.8}(k)^{.6}(C/\mu)^{.4}(1/x^{.2})\Delta T \qquad (2\text{-}14)$$

Eq. (2-14) is in the $q\{\Delta T\}$ form required for analysis in the new heat transfer. Notice that this form is quite transparent. It readily reveals most of the parameters which have designer importance, and it readily reveals their relationship to each other.

(Eq. (2-14) is not completely in the form of the new heat transfer. We will complete the conversion of Eq. (2-13) in a later chapter.)

CHAPTER 3

HOW TO SOLVE PROBLEMS USING THE NEW HEAT TRANSFER

The solution of problems is extremely simple using the new heat transfer, even if the problems involve highly nonlinear behavior. The simplicity results because the new heat transfer uses the same straightforward mathematics used in pure mathematics. (The old heat transfer uses a peculiar type of mathematics which is roundabout and bears little resemblance to pure mathematics.)

The first part of this chapter discusses the solution of heat transfer problems in a general way. We formulate the generic problem which is common to both old and new heat transfers, and then we solve the problem both ways.

In the second part of this chapter, the new heat transfer is used to solve several textbook type problems involving proportional and nonlinear heat transfer behavior. Problems (3-1) to (3-5) are stated and the required information given within the framework of the new heat transfer--ie in terms of thermal behavior and *not* in terms of heat transfer coefficients. The problems are solved using the new heat transfer, and it will be seen that the solution of every problem is quite simple and obviously correct.

Problems (3-6) to (3-8) are the same as Problems (3-2) to (3-4) except that they are stated in the language of the old heat transfer. The reader is asked to solve Problems (3-6) to (3-8) using the old heat transfer--ie using heat transfer coefficients. Be forewarned that these problems involve nonlinear behavior, and therefore you will have considerable difficulty solving them with the old heat transfer. So much difficulty that you will probably quit before you obtain the correct and complete answers.

A GENERIC PROBLEM AND THE OLD/NEW WAY SOLUTIONS

A generic problem for the heat transfer analyst is:

- Determine the local heat flux in the heat transfer system shown in Fig. (3-1).
- Determine the total heat flow rate by integrating the local heat flux over the entire surface.

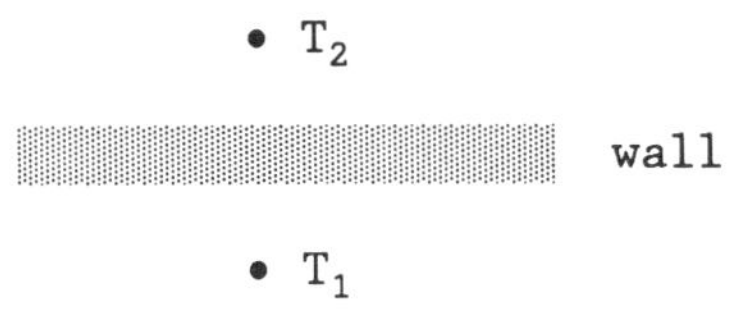

FIGURE (3-1) THERMAL SYSTEM

This problem is generic in both heat transfers. Its solution requires a specification of the overall temperature difference and a description of the heat transfer behavior of each thermal component--ie a description of the thermal behavior of the wall and the two interfaces in Fig. (3-1).

As indicated in the next several sections, the description of component behavior and the determination of local heat flux differ greatly in the two heat transfers. However, there is no difference in the integration of local heat flux.

DESCRIPTION OF HEAT TRANSFER BEHAVIOR--OLD HEAT TRANSFER

In the old heat transfer, the heat transfer behavior of thermal components is described in terms of the coefficients h, k, and ϵ. Typical examples are:

$$k = 8.5 \tag{3-1}$$

$$h = 13 \tag{3-2}$$

$$h = 140(1 + .012T_b)V^{.8}/D^{.2} \tag{3-3}$$

$$h = 0.21\ \Delta T^{.33} \tag{3-4}$$

$$Nu = .023\ Re^{.8}\ Pr^{.4} \tag{3-5}$$

$$h = 0.35\ \Delta T^{.25} \tag{3-6}$$

$$\epsilon = 0.91 \tag{3-7}$$

$$k = 0.033 + .00002T \tag{3-8}$$

graph of h vs ΔT (3-9)

DESCRIPTION OF HEAT TRANSFER BEHAVIOR--NEW HEAT TRANSFER

In the new heat transfer, the heat transfer behavior of thermal components is described in terms of thermal behavior--ie in the generic form q{TDF}. Examples of specific forms are:

$$q = 8.5\ \frac{dT}{dx} \simeq 8.5\ \frac{\Delta T}{\Delta x} \tag{3-1a}$$

$$q = 13\ \Delta T \tag{3-2a}$$

$$q = 140(2 + .012T_b)V^{.8}\Delta T/D^{.2} \tag{3-3a}$$

$$q = 0.21\ \Delta T^{1.33} \tag{3-4a}$$

$$q = .023\ G^{.8}a_0^{.6}(C/\mu)^{.4}(1/x^{.2})\Delta T \tag{3-5a}$$

$$q = 0.35\ \Delta T^{1.25} \qquad (3\text{-}6a)$$

$$q = 0.91\ \sigma\ T^4 \qquad (3\text{-}7a)$$

$$q = (.033 + .00002T)(dT/dx) \qquad (3\text{-}8a)$$

graph of q vs ΔT (3-9a)

Notice that Eq. (3-1a) is identical to Eq. (3-1), and similarly for the other expressions. The *only* difference between the old and the new expressions is language. Expressions (3-1) to (3-9) are written in the language of coefficients--of h, k, and ϵ. Expressions (3-1a) to (3-9a) are written in the language of thermal behavior--of q{TDF}--a language which has nothing to do with parameters of the intellect--nothing to do with h, or k, or ϵ or Nu or Re or Gr or Ra or . . . .

SOLUTION OF GENERIC PROBLEM--OLD HEAT TRANSFER

Using the old heat transfer to analyze the system shown in Fig. (3-1), the given information would consist of:

- T_1 and T_2.
- Thermal conductivity of the wall. Eq. (3-1) is typical.
- An h correlation for each interface. Eqs.(3-2) to (3-6) are typical h correlations.
- Miscellaneous information required to evaluate the parameters in the h correlations.

Given this information, the local heat flux is obtained from Eqs. (3-10) and (3-11):

$$U = \frac{1}{1/h_1 + 1/h_2 + t/k} \qquad (3\text{-}10)$$

$$q = U\ \Delta T_{total} \qquad (3\text{-}11)$$

SOLUTION OF GENERIC PROBLEM--NEW HEAT TRANSFER

Using the new heat transfer to analyze the system in Fig. (3-1), the given information would consist of:

- T_1 and T_2.
- Thermal behavior of the wall. Eq. (3-1a) is typical.
- A thermal behavior correlation for each interface. Eqs. (3-2a) to (3-6a) are typical.
- Miscellaneous information required to evaluate the parameters in the thermal behavior correlations.

Given this information, the local heat flux is obtained from Eq. (3-10a):

$$\Delta T_{total} = \Delta T_1 + \Delta T_{wall} + \Delta T_2 \qquad (3\text{-}10a)$$

This equation is solved as follows:

- Rearrange each given thermal behavior expression into the form $\Delta T = f\{q\}$.
- Substitute the corresponding $f\{q\}$ for each ΔT on the right side of Eq. (3-10a).
- Note that q is continuous and therefore $q_1 = q_w = q_2$.
- Solve for q.

This new way solution may seem much less desirable than the old way, particularly if you have used the old way for many years. But as the problems ahead demonstrate, the solution of problems involving nonlinear heat transfer behavior is much simpler using the new methodology. And when you get accustomed to the new way, you will find that the solution of problems involving only proportional behavior is equally simple using the old and the new way.

THE PURPOSE OF PROBLEMS (3-1) TO (3-8)

Problems (3-1) to (3-5) illustrate the determination of local heat flux using the methodology of the new heat transfer. Problems (3-6) to (3-8) restate Problems (3-2) to (3-4) within the framework of the old heat transfer. They are intended to be solved by the reader using the old heat transfer--ie using heat transfer coefficients.

Problems (3-1), (3-2), and (3-5) are typical of problems generally presented in texts on the old heat transfer. Problems (3-1) and (3-5) involve thermal behavior which is strictly proportional. Problem (3-2) includes an interface which exhibits moderately nonlinear behavior (as in free convection heat transfer).

It should be noticed that, in a practical sense, the new way solutions of Problems (3-1) and (3-5) do not differ greatly from old way solutions of similar problems. That is because, when a problem involves only behavior which is strictly proportional, "$q = h\Delta T$" does not merely define h--it also describes $q\{\Delta T\}$.

Problem (3-5) illustrates the integration of local heat flux in equipment which involves heat transfer behavior which is strictly proportional and does not vary with location. In this simple type of equipment, the old/new integration is expressed in the following closed form:

$$Q = \int q\ dA = \int U\Delta T\ dA = U\ A\ \Delta T_{LM} \qquad \text{(old way)}$$

$$Q = \int q\ dA = q_{LM}\ A \qquad \text{(new way)}$$

Problem (3-2) contains a moderately nonlinear interface, and this degree of nonlinearity is succesfully dealt with in the old heat transfer. However, notice that the new way solution is simple and direct, whereas the old way solution in Problem (3-6) is indirect because h makes it impossible to separate the variables q and ΔT.

Problems (3-3) and (3-4) bear no resemblance to problems presented in texts on the old heat transfer. The behavioral analysis of equipment containing highly nonlinear interfaces is not within the knowledge envelope of the old heat transfer. Yet it should be noted that the new way analysis of Problems (3-3) and (3-4) is simple and straightforward, and the results are obviously correct.

The reader is encouraged to make a serious effort to solve Problems (3-6) to (3-8) using the old heat transfer. The result will be the recognition that the old heat transfer does not deal effectively with nonlinear phenomena.

Problems (3-7) and (3-8) are so difficult to solve using the old heat transfer that it is unlikely that *any* reader will obtain the correct and complete solution to either problem using the old heat transfer--ie using "heat transfer coefficients". And neither of these problems deals with the most difficult form of nonlinearity--a form we will deal with in a very simple way in Chapter 36--using the new heat transfer.

The solution of Problems (3-3) and (3-4) has nothing to do with h, whereas the reader's solution of Problems (3-7) and (3-8) is intended to be based on h. This one difference in methodology is solely responsible for the fact that the new way solutions are quite simple, whereas the old way solutions border on the impossible.

PROBLEM (3-1) PROPORTIONAL THERMAL BEHAVIOR

Determine the heat flux in the thermal system shown in Fig. (3-2).

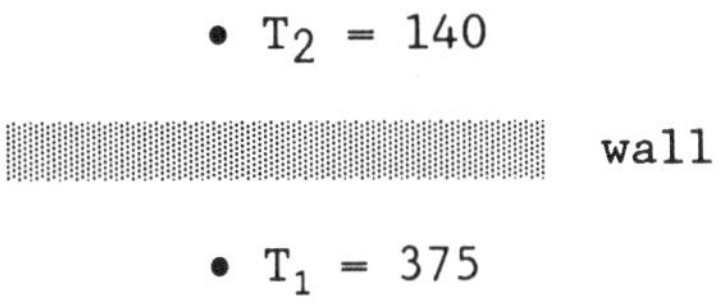

FIGURE (3-2) THERMAL SYSTEM IN PROBLEM (3-1)

INPUT TO PROBLEM (3-1)

The thermal behavior of both interfaces in Fig. (3-2) is given by Eq. (3-5a). When the fluid and system properties are substituted into Eq (3-5a), Equations (3-12) and (3-13) result:

$$q_1 = 25.7\ \Delta T_1 \tag{3-12}$$

$$q_2 = 43.5\ \Delta T_2 \tag{3-13}$$

The thermal behavior of the wall in Fig. (3-2) is given by Eq. (3-1a). The wall thickness is 0.05 ft.

$$q_{wall} = 8.5\ \Delta T/\Delta x = (8.5/.05)\Delta T_{wall} = 170\ \Delta T_{wall}$$

ANALYSIS OF PROBLEM (3-1)

The first step is to recall the keystone of the new heat transfer,

$$\Delta T_{total} = \sum \Delta T_i \tag{3-14}$$

$$\Delta T_{total} = \Delta T_1 + \Delta T_w + \Delta T_2 \tag{3-15}$$

From the given information and the observation that q is independent of location, we may write:

$$\Delta T_1 = q/25.7 \tag{3-16}$$

$$\Delta T_{wall} = q/170 \tag{3-17}$$

$$\Delta T_2 = q/43.5 \tag{3-18}$$

Substituting these expressions into Eq. (3-15) and noting that q is continuous gives:

$$(375 - 140) = q/25.7 + q/170 + q/43.5 \tag{3-19}$$

ANSWER TO PROBLEM (3-1)

The solution of Eq. (3-19) gives $q = 3467$. (Notice that we do not give the dimensions in explicit form because, in the new heat transfer, the dimensions are implicit in the symbols.)

DISCUSSION OF PROBLEM (3-1)

Problem (3-1) involves strictly proportional behavior, and problems involving this simple type of behavior are solved directly in both heat transfers. But when the behavior is not proportional, problems can be solved directly only if one uses the new heat transfer--as demonstrated in the next few problems which involve nonproportional behavior.

PROBLEM (3-2) MODERATELY NONLINEAR THERMAL BEHAVIOR

Determine the heat flux in the thermal system shown in Fig. (3-3):

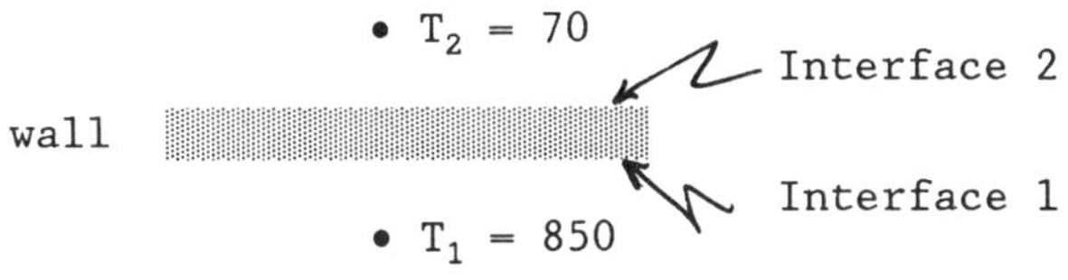

FIGURE (3-3) THERMAL SYSTEM IN PROBLEM (3-2)

INPUT TO PROBLEM (3-2)

The thermal behavior of interface 1 is given by Eq. (3-5a). When the fluid and system properties are substituted into Eq. (3-5a), Eq. (3-20) results:

$$q_1 = 13.3\ \Delta T_1 \tag{3-20}$$

The thermal behavior of interface 2 is given by Eq. (3-4a).

$$q = 0.21\ \Delta T^{1.33} \tag{3-4a}$$

The thermal behavior of the wall is given by Eq. (3-1a). The wall thickness is 0.08 ft.

$$q_{wall} = 8.5\ \Delta T/\Delta x = (8.5/.08)\Delta T_{wall} = 106.3\ \Delta T_{wall}$$

ANALYSIS OF PROBLEM (3-2)

The first step is to recall the keystone of the new heat transfer:

$$\Delta T_{total} = \Delta T_1 + \Delta T_w + \Delta T_2 \qquad (3\text{-}21)$$

From the given information and the observation that q is continuous, we may write:

$$\Delta T_1 = q/13.3 \qquad (3\text{-}22)$$

$$\Delta T_w = q/106.3 \qquad (3\text{-}23)$$

$$\Delta T_2 = (q/.41)^{.75} \qquad (3\text{-}24)$$

Substituting into Eq. (3-21) gives:

$$(850\text{-}70) = q/13.3 + q/106.3 + (q/.41)^{.75} \qquad (3\text{-}25)$$

ANSWER TO PROBLEM (3-2)

The solution of Eq. (3-25) gives *q = 2090*.

DISCUSSION

Note that the new way solution of this problem involving nonlinear behavior is *exactly the same* as the new way solution of Problem (3-1) which involves proportional behavior. The only difference in the analysis of the two problems is that we must solve a nonlinear equation instead of a proportional one--Eq. (3-25) instead of Eq. (3-19).

The solution of this simple problem is considerably more difficult using the old heat transfer--ie using h. The difficulty is demonstrated in Problem (3-6) which repeats the above problem using old heat transfer methodology.

PROBLEM (3-3) HIGHLY NONLINEAR THERMAL BEHAVIOR (ANALYTICAL SOLUTION)

Determine the heat flux in the thermal system shown in Fig. (3-4).

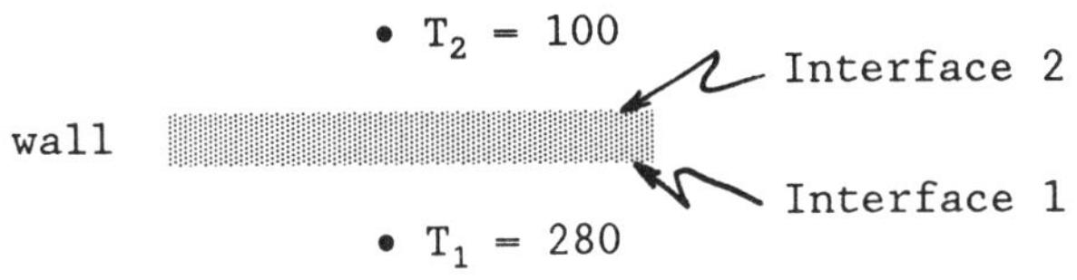

FIGURE (3-4) THERMAL SYSTEM, PROBLEM (3-3)

INPUT TO PROBLEM (3-3)

The thermal behavior of interface 1 is given by Eq. (3-5a). When the fluid and system properties are substituted into Eq. (3-5a), Eq. (3-26) results:

$$q_1 = 6300\ \Delta T_1 \qquad (3\text{-}26)$$

The thermal behavior of the wall is described by Eq.(3-1a). The wall thickness is .00759 ft. Therefore

$$q_{wall} = 8.5\ \Delta T/\Delta x = 1120\ \Delta T_{wall} \qquad (3\text{-}27)$$

The thermal behavior of interface 2 is given by Eq. (3-28):

$$q_2 = 10000\Delta T_2 - 250\Delta T_2^{\,2} + 15.5\Delta T_2^{\,2.5} \qquad (3\text{-}28)$$

ANALYSIS OF PROBLEM (3-3)

The analysis of this problem would be identical to the preceding problems if we could convert Eq. (3-28) from the form $q\{\Delta T\}$ to the form $\Delta T\{q\}$. Since this conversion is not readily apparent, we can avoid it by solving the problem in a slightly different way. This different way involves obtaining an expression for the heat flux into Interface 2, and equating the heat flux into this interface and the heat flux out. This involves noting that

$$(T_1 - T_{i,2}) = \Delta T_1 + \Delta T_{wall} \tag{3-29}$$

where $T_{i,2}$ refers to the wall temperature at Interface 2. Combining the given information with Eq. (3-29) gives:

$$(280 - T_{i,2}) = q/6300 + q/1120 \tag{3-30}$$

$$q = 951(280 - T_{i,2}) \tag{3-31}$$

We obtain Eq. (3-32a) by inspection of Fig. (3-4); Eq. (3-33) by combining (3-31) and (3-32c); Eq. (3-34) by combining (3-28) and (3-33).

$$(T_1 - T_{i,2}) = (T_1 - T_2 - \Delta T_2) \tag{3-32a}$$

$$(280 - T_{i,2}) = (280 - 100 - \Delta T_2) \tag{3-32b}$$

$$T_{i,2} = 180 + \Delta T_2 \tag{3-32c}$$

$$q = 951(180 - \Delta T_2) \tag{3-33}$$

$$951(180 - \Delta T_2) = 10000\Delta T_2 - 250\Delta T_2^2 + 15.5\Delta T_2^{2.5} \tag{3-34}$$

The solution of Eq. (3-34) gives 3 values of ΔT_2:

$$\Delta T_2 = 27 \text{ or } 74 \text{ or } 125$$

We shall see in the later chapter on thermal stability that operation at ΔT_2=74 is thermally unstable. The end result of the instability is that the system would not continue to operate at ΔT_2=74. Even without an outside disturbance, operation would shift to one of the other two values where steady-state operation is possible.

ANSWER TO PROBLEM (3-3)

The system described in Problem (3-3) operates at a steady-state heat flux of 147000 or 52000 Btu/hrft^{2}F. (This result is obtained from the solution of Eq. (3-28) for ΔT_2 values of 27 and 125.) The given information is not sufficient to determine a unique solution.

As we shall see in a later chapter on thermal stability, a description of the prior operating history would be needed in order to determine the unique solution to this problem.

PROBLEM (3-4) HIGHLY NONLINEAR THERMAL BEHAVIOR (GRAPHICAL SOLUTION)

Determine the steady-state heat flux in the thermal system shown in Fig. (3-5).

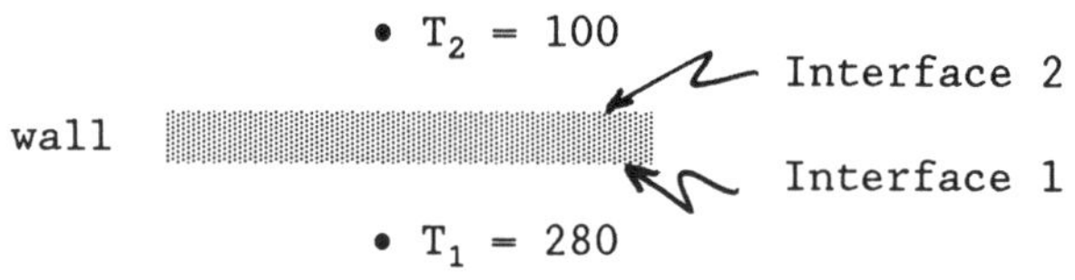

FIGURE (3-5) THERMAL SYSTEM, PROBLEM (3-4)

INPUT TO PROBLEM (3-4)

The thermal behavior of Interface 1 is given by Eq. (3-5a). The wall thermal behavior is given by Eq. (3-1a). When the system and fluid properties are substituted into these equations, Eqs. (3-34) and (3-35) result:

$$q_1 = 6300\ \Delta T_1 \tag{3-34}$$

$$q_{wall} = 8.5\ \Delta T/\Delta x = 1120\ \Delta T_{wall} \tag{3-35}$$

Fig. (3-6) describes the thermal behavior of Interface 2.

ANALYSIS OF PROBLEM (3-4)

This problem is identical to Problem (3-3) except that the behavior of interface 2 is described graphically rather than analytically. We can solve the problem in one of two ways:

- *Analytically* by determining an analytical expression which approximates the curve in Fig. (3-6), then solving the problem in the manner of Problem (3-3).

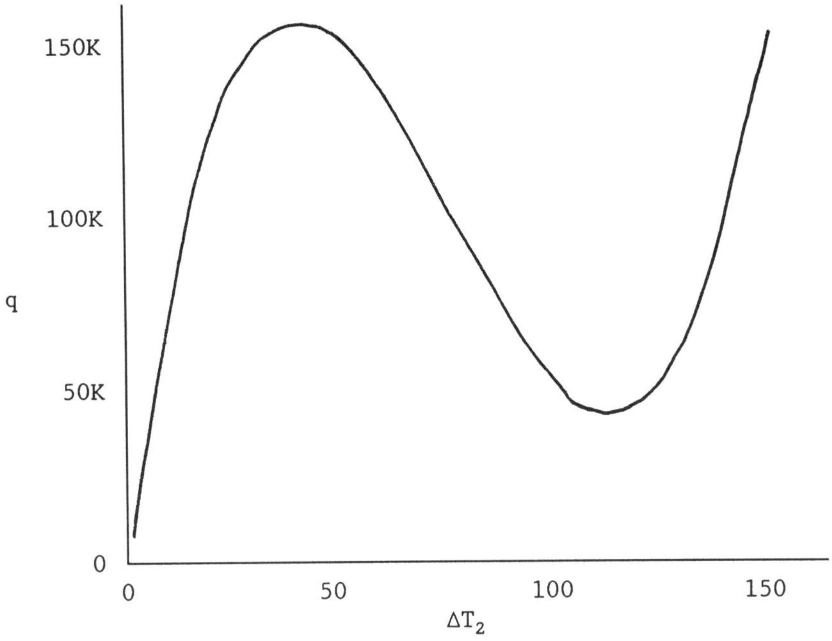

FIGURE (3-6) THERMAL BEHAVIOR OF INTERFACE 2, PROBLEM (3-4)

- *Graphically* by determining an expression for the heat flux into Interface 2, then plotting this expression on Fig. (3-6). Since the heat flux into the interface is equal to the heat flux out of the interface only at intersections, and since heat flux is continuous, the intersections are the only potential operating points.

  ("Potential" operating points in that the system may be thermally unstable at one or more intersections. Instability would result in a tendency to move away from the intersection, or would result in undamped oscillatory operation in the region of the intersection. Which effect would actually result depends on the thermal behavior of the components, as we shall see in a later chapter.)

The determination of analytical expressions which describe highly nonlinear curves is a time-consuming process which often requires considerable approximation. Therefore graphical solutions are usually simpler and more accurate when dealing with highly nonlinear behavior.

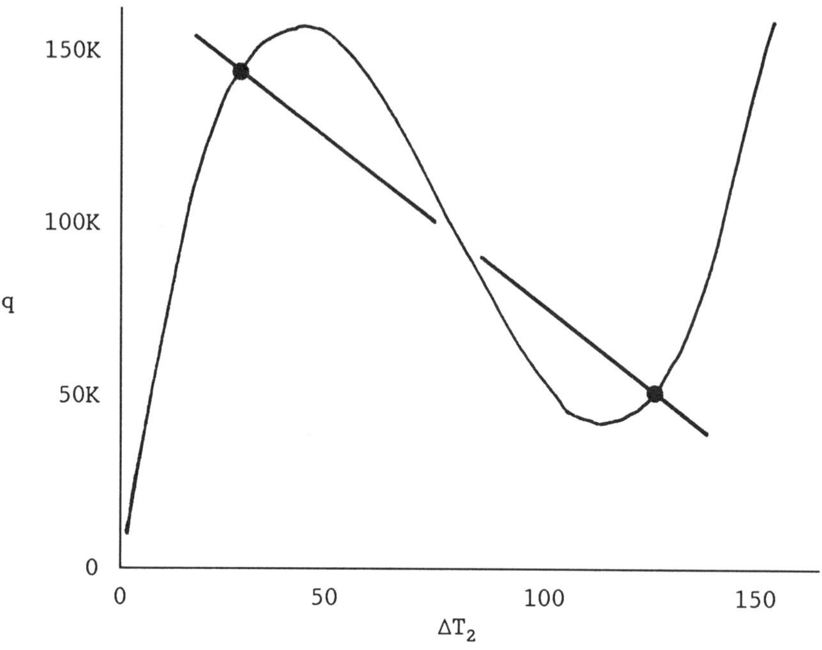

FIGURE (3-7) GRAPHICAL SOLUTION OF PROBLEM (3-4)

In order to illustrate the method, we will solve Problem (3-4) graphically. But it should be noted that the graphical solution of Problem (3-4) is exactly analogous to the analytical solution of Problem (3-3). The two methods give exactly the same result, and are equally rigorous.

The solution of this problem involves the graphical solution of two simultaneous equations. Fig. (3-6) is the graphical expression of one of the equations. Eq. (3-36) is the analytical expression of the other, and is obtained by combining Eqs. (3-34) and (3-35). (Notice that Eq. (3-36) was Eq. (3-33) in the preceding problem.)

$$q = 951(180 - \Delta T_2) \tag{3-36}$$

We solve for q and ΔT_2 by plotting Eq. (3-36) on Fig. (3-6), and noting that intersections are the only potential steady-state operating points. Fig. (3-7) is the result of plotting Eq. (3-36) on Fig. (3-6).

ANSWER TO PROBLEM (3-4)

The intersections in Figure (3-7) indicate that the steady-state heat flux in the given system is either 147000 or 52000 Btu/hrft^2. The given information is not sufficient to determine a unique solution. We would need to know the prior operating history of the equipment in order to determine the unique solution to the problem.

(The middle intersection is not a potential operating point because, as we shall see in the chapter on thermal stability, the system is thermally unstable at this intersection. The shape of the curves in Fig. (3-6) indicates that the system would tend to move away from the middle intersection until it reached one of the stable intersections.)

DISCUSSION OF PROBLEMS (3-3) AND (3-4)

Problems dealing with highly nonlinear thermal behavior (such as Problems (3-3) and (3-4)) are not found in texts on the old heat transfer. The old heat transfer simply can not cope with highly nonlinear behavior.

For example, the highly nonlinear thermal behavior of pool boiling interfaces is often published, but the behavioral analysis of equipment which contains such interfaces is outside the knowledge envelope of the old heat transfer. This is evidenced by the simplistic description of pool boiler behavior given in current heat transfer texts.

In the old heat transfer view, thermal stability is assured in "temperature controlled" heat exchangers such as the equipment in Problems (3-3) and (3-4). Recall the statement by Rohsenow (1963):

> *With condensing vapor as the heat source on one side of a wall, any point on the entire (pool boiling) curve can be reached under stable conditions.*

Notice that the equipment in Problems (3-3) and (3-4) closely resembles a pool boiler with condensing vapor as the heat source, and yet it is *NOT* able to reach any point on the curve in Fig. (3-6) under stable conditions.

In the new heat transfer, we deal effectively with highly nonlinear thermal behavior in a very simple way, as demonstrated by Problems (3-3) and (3-4). In order to make the analysis as simple as possible, we often perform the analysis graphically as in Problem (3-4).

Note that the analysis of Problems (3-3) and (3-4) involves "uncoupling" the system at Interface 2. In other words, we mentally/analytically separate the system into two smaller systems:

- The heat source system bounded by T_1 and $T_{i,2}$.

- The heat sink system bounded by $T_{i,2}$ and T_2.

We analyze the heat source system to determine how heat flow *INTO* Interface 2 responds to the ΔT across the interface. Eq. (3-36) is the result of that analysis. Fig. (3-6) describes the heat sink system--it describes how the heat flow *OUT OF* the interface responds to the temperature difference across the interface.

We note that q is continuous, and therefore if we "couple" the source and the sink systems together, the heat flow into the sink must equal the heat flow out of the source. We determine the result of coupling the source and sink by plotting the thermal behavior of the source and sink on the same graph, and noting that q is continuous only at intersections. Therefore only intersections are potential operating points. ("Potential" because one or more intersections may be thermally unstable and would not represent a steady-state operating point.)

FLUID FLOW ANALOG OF PROBLEM (3-4)

It is instructive to note that a certain fluid flow problem is exactly analogous to Problem (3-4). It is the problem of determining the fluid flow rate which would result if a given pump were installed in a given piping system.

In the problem, the hydraulic characteristic of the piping system is given in the form ΔP vs W, and the pump head curve is also given in the form ΔP vs W. We determine the fluid flow rate by plotting the pump head curve and the system hydraulic characteristic on the same graph, noting that the intersection(s) represent potential steady-state operating conditions. Fig. (3-8) illustrates the method.

In the fluid flow analog, the pump is the uncoupled fluid "source", and the piping system is the uncoupled fluid "sink". The interface "surface" is the pump connections to the piping system. We determine the result of coupling the source and the sink by plotting their separate hydraulic behavior on a common graph and noting that intersection(s) represent potential steady-state operating conditions. As in the heat transfer problem, we must stipulate "potential" because intersections may be unstable.

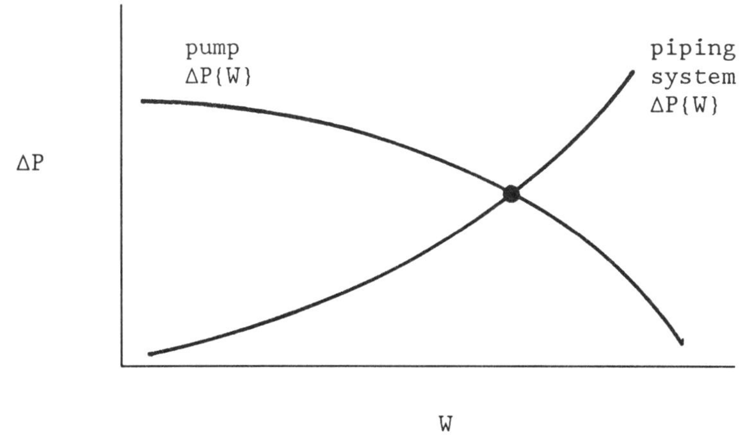

FIGURE (3-8) FLUID FLOW ANALOG OF PROBLEM (3-4)

PROBLEM (3-5) HEAT FLOW INTEGRATION IN SYSTEMS WHERE THE THERMAL BEHAVIOR IS PROPORTIONAL AND INDEPENDENT OF LOCATION

The countercurrent, one pass, shell and tube heat exchanger shown in Fig. (3-9) is being designed to heat 1200 lbs/hr of Fluid A from 200 to 360 °F. Given the information below, what heat transfer area (A, ft^2) is required?

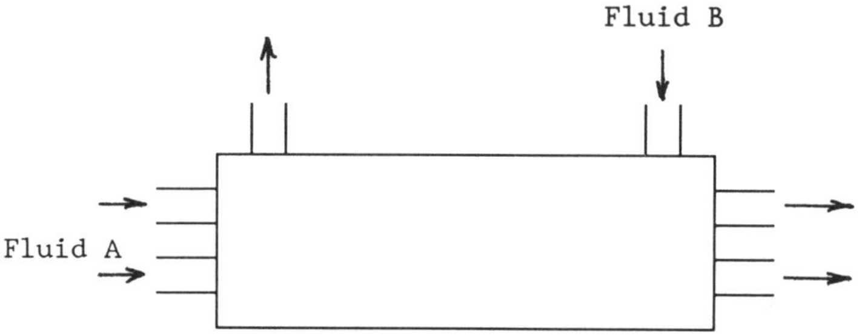

FIGURE (3-9) HEAT EXCHANGER, PROBLEM (3-5)

INPUT TO PROBLEM (3-5)

The heating fluid is 1600 lbs/hr of Fluid B at T = 400. A heat balance indicates that the design heat flow rate (Q, Btu/hr) is 115,000 Btu/hr, and that the exit temperature of Fluid B is 324 F. Therefore the TDF will be 40 F at the hot end of the exchanger, and 124 F at the cold end.

The thermal behavior of the tube wall and interfaces are independent of location. The thermal behavior of both interfaces is described by Eq. (3-5a). The wall thermal behavior is described by Eq. (3-1a). When the system and fluid properties are substituted in these equations, the following equations result:

$$q_A = 480\ \Delta T_A \tag{3-37}$$

$$q_B = 620\ \Delta T_B \tag{3-38}$$

$$q_{wall} = 2850\ \Delta T_{wall} \tag{3-39}$$

ANALYSIS OF PROBLEM (3-5)

The solution of Problem (3-5) involves combining Eqs. (3-37) to (3-39) into a single equation of the form $q\{\Delta T_{total}\}$, then determining the total heat flow rate by integration. Since the thermal behavior is proportional and independent of location, integration is the simple matter described in Eq. (3-40):

$$Q = \int q \, dA = q_{LM} A \qquad (3\text{-}40)$$

$$q_{LM} \equiv q\{\Delta T_{total} = \Delta T_{LM}\} \qquad (3\text{-}41)$$

As in the previous problems, we combine Eqs. (3-37) to (3-39) by recalling the keystone of the new heat transfer:

$$\Delta T_{total} = \Delta T_1 + \Delta T_2 + \Delta T_{wall} \qquad (3\text{-}42)$$

$$\Delta T_{total} = q/480 + q/620 + q/2850 \qquad (3\text{-}43)$$

$$q = 247 \, \Delta T_{total} \qquad (3\text{-}44)$$

$$\Delta T_{LM} = (124 - 40)/\ln(124/40) = 74.2 \qquad (3\text{-}45)$$

$$q_{LM} = 247(74.2) = 18300 \qquad (3\text{-}46)$$

$$Q = 115000 = q_{LM} A = 18300A \qquad (3\text{-}47)$$

$$A = 6.28 \qquad (3\text{-}48)$$

ANSWER TO PROBLEM (3-5)

The required heat exchanger surface area is 6.28 ft^2.

DISCUSSION OF PROBLEM (3-5)

The proof that the integration in Eq. (3-40) is correct is the same as the old heat transfer proof that integration yields the result in Eq. (3-50) if "U" is independent of both ΔT and location in the equipment:

$$Q = \int q \; dA = U \; A \; \Delta T_{LM} \qquad (3\text{-}50)$$

Notice that

$$U \; \Delta T_{LM} = q\{\Delta T_{total} = \Delta T_{LM}\} \equiv q_{LM} \qquad (3\text{-}51)$$

and therefore the integration in Eq. (3-40) is identical to the integration in (3-50). The only difference is that Eq. (3-40) is written in the language of the new heat transfer, and Eq. (3-50) is written in the language of the old heat transfer.

It is important to note that integration is accomplished in the simple manner of Eqs. (3-40) and (3-50) *ONLY* if the thermal behavior is everywhere proportional and independent of location. If either condition is not satisfied, the integration would probably be computer aided.

PROBLEM (3-6) REPEAT PROBLEM (3-2) USING THE OLD HEAT TRANSFER

Determine the heat flux in the thermal system shown in Fig. (3-10).

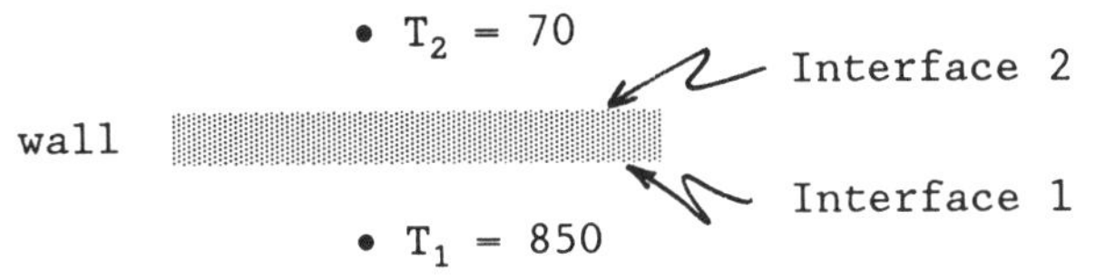

FIGURE (3-10) THERMAL SYSTEM IN PROBLEM (3-6)

INPUT TO PROBLEM (3-6)

The heat transfer coefficient at interface 1 is given by Eq.(3-5). When the fluid and system properties are substituted into Eq. (3-5), Eq. (3-52) results:

$$h_1 = 13.3 \text{ Btu/hr ft} \tag{3-52}$$

The h at interface 2 is given by Eq. (3-4):

$$h_2 = 0.21 \, \Delta T_2^{.33} \tag{3-4}$$

The thermal conductivity of the wall is 8.5 Btu/hr ft^2. The wall thickness is 0.08 ft.

ANALYSIS OF PROBLEM (3-6)

To be completed by the reader.

Note that the solution of this problem involving moderately nonlinear thermal behavior is difficult using h and the old heat transfer, whereas the solution was simple using the new heat transfer.

PROBLEM (3-7) REPEAT PROBLEM (3-3) USING THE OLD HEAT TRANSFER

Determine the steady-state heat flux in the thermal system shown in Fig. (3-11).

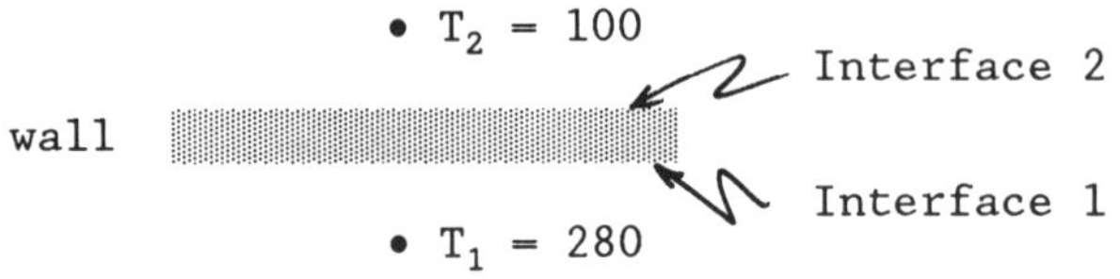

FIGURE (3-11) THERMAL SYSTEM, PROBLEM (3-7)

INPUT TO PROBLEM (3-7)

h_1 = 6300 Btu/hr ft^2

t_{wall} = .00759 ft.

k_{wall} = 8.5 Btu/hr ft^2

$$h_2 = 10000 - 250\Delta T_2 + 15.5\Delta T^{1.5} \qquad (3\text{-}53)$$

ANALYSIS OF PROBLEM (3-7)

To be completed by the reader.

Notice that it is extremely difficult to obtain the correct and complete solution to this problem using h and the old heat transfer. And recall that the solution of this same problem was simple and straightforward using the new heat transfer.

PROBLEM (3-8) REPEAT PROBLEM (3-4) USING THE OLD HEAT TRANSFER

Determine the steady-state heat flux in the thermal system shown in Fig. (3-12).

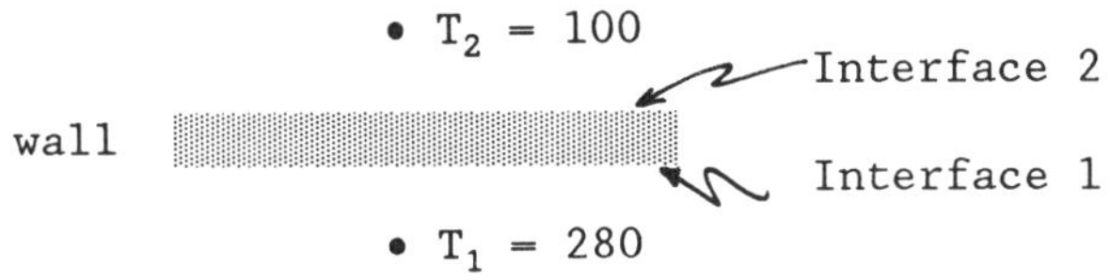

FIGURE (3-12) THERMAL SYSTEM, PROBLEM (3-8)

INPUT TO PROBLEM (3-8)

h_1 = 6300 Btu/hr ft^2

t_{wall} = .00759 ft.

k_{wall} = 8.5 Btu/hr ft^2

h_2 = as described in Fig. (3-13)

ANALYSIS OF PROBLEM (3-8)

To be completed by the reader.

Notice that it is extremely difficult to obtain the correct and complete solution to this problem using h and the old heat transfer. And recall that the solution of this problem was quite simple and straightforward using the new heat transfer.

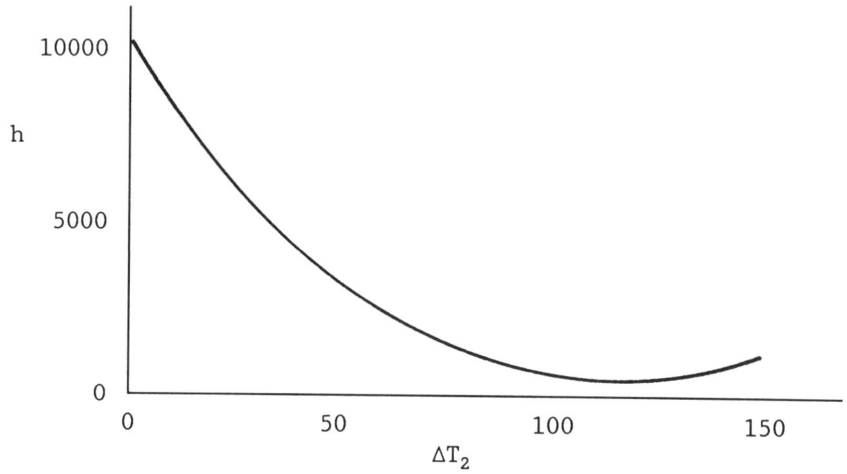

FIGURE (3-13) **HEAT TRANSFER COEFFICIENT AT INTERFACE 2 IN PROBLEM (3-8)**

TABLE (3-1)

COORDINATES FOR FIGURE (3-13)

ΔT	h
0	10000
10	7990
20	6386
30	5047
40	3921
50	2980
60	2204
70	1578
80	1091
90	734
100	500
110	382
120	375
130	475
140	676
150	975

CHAPTER 4

WHAT EXACTLY IS h?

Texts on today's old heat transfer describe how to use h, but do NOT describe exactly what h is. For example, there is a graphical analog of h which is useful in describing h. Yet heat transfer texts do not use the graphical analog to promote an understanding of h.

Heat transfer texts logically omit the graphical analog because, in today's old heat transfer, h is viewed as a fundamental parameter--a parameter which truly exists and can be measured. Parameters which are truly fundamental do not have graphical analogs. There is no graphical analog of heat flux--or temperature--or time--or distance. If h were a fundamental parameter of Nature, it too would have no graphical analog.

This chapter has one purpose--to describe *exactly* what h is. It is necessary to understand exactly what h is in order to appreciate:

- What's wrong with h?
- Why must h be abandoned?
- Why is the thermal behavior concept better than the heat transfer coefficient concept?

A SIMPLE AND REVEALING QUIZ ON h

The quiz on the next page is designed to appraise your understanding of h. Many persons have taken this quiz, and even though they were quite competent in the old heat transfer, most of them got a *zero* on the quiz in what I consider a reasonable length of time (5 minutes). When the quiz was taken without time limit, the average grade was approximately 25%.

If you really know what h is, you will have no difficulty answering the questions, and it will take you less than 5 minutes to complete the quiz. Take as long as you like, then put the quiz aside. After you have read this chapter which discusses the significance of both the questions and the answers, grade your solution and answer the final question:

"Did the person who took this quiz really know what h is?"

THE QUIZ ON h

1. Given that $q = h\ \Delta T$, what is $dq/d\Delta T$?

2. Velocity is the derivative of distance with respect to time. What exactly is h?

3. Eq. (4-1) is a complicated form of a very simple equation. Transform Eq. (4-1) into the simple form.

$$U = \frac{1}{1/h_1 + 1/h_2 + t/k} \qquad (4\text{-}1)$$

4. The slope of the dashed line in Fig. (4-1) is the graphical analog of velocity at point A. Draw the graphical analog of h at point B in Figure (4-2).

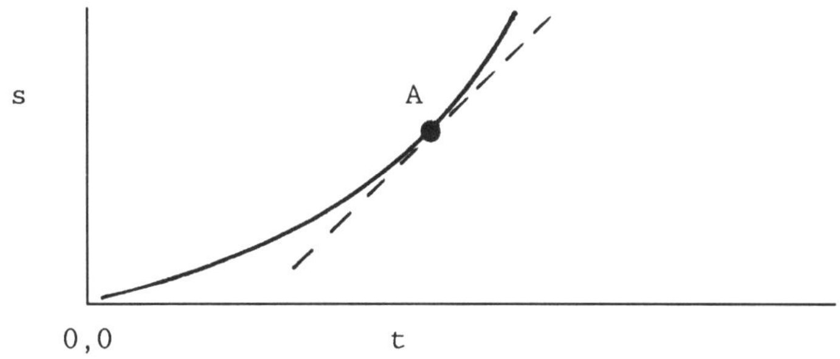

FIGURE (4-1) Graphical analog of velocity

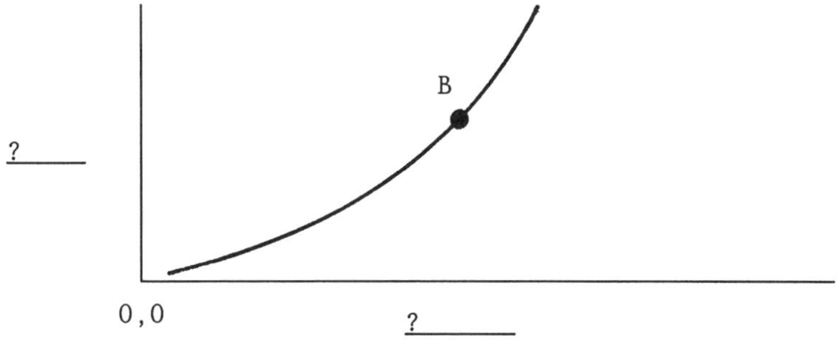

FIGURE (4-2) Graphical analog of h
(to be completed by the reader)

WHAT IS TODAY'S h?

The h concept was conceived by Fourier (1822), but it was revised long ago, and today's h little resembles the h conceived by Fourier. Today's (1989) h is accurately described by *all* of the following:

- h is not a constant in the equation $q = h\Delta T$.
  h is a *variable* in the equation.

- h is not a fundamental parameter.
  h is a *contrived* parameter.

- h is *not* found in Nature.
  h is found only in the intellect.
  h does *not* actually exist!

- Since h does not exist, h can *not* be measured.

- Experimenters measure q and ΔT and then *generate/invent/create* h by dividing q by ΔT.

- h is nothing more than a shorthand way of writing "the ratio of heat flux to boundary layer temperature difference".

- h can *always* be replaced by $(q/\Delta T)$ because h *IS* $(q/\Delta T)$.

TODAY'S TEXTBOOK DESCRIPTION OF h

The first few pages of today's heat transfer texts state that h is defined by

$$q = h \, \Delta T \qquad (4\text{-}2)$$

The first few pages also point out that Eq. (4-2) is a definition of h and *NOT* a phenomenological description of convective heat transfer. In other words, Eq. (4-2) says *NOTHING* about functionality, even though it appears to state that q is proportional to ΔT. q is *not* generally proportional to ΔT, and therefore the value of "h" often depends on the value of ΔT.

(Note that the phrase "Eq. (4-2) is a definition of h" is identical to the phrase "Eq. (4-2) states that h is an alias for the ratio of q to ΔT".)

After the first few pages, texts concentrate on h in such a way that we lose sight of the fact that h is nothing other than an alias for the ratio (q/ΔT). The first few pages are soon forgotten, and h gradually assumes a character of its own. We forget that h is simply a name we have given to a dimensional group, and we begin to think of h as a parameter of Nature--a parameter which really exists.

It would be far better and more revealing to write Eq. (4-2) in the form:

$$h \equiv q/\Delta T \qquad (4\text{-}3)$$

The form of Eq. (4-3) is preferable because it does not imply proportionality. It obviously says nothing about the relationship between q and ΔT. It says exactly what is meant, and nothing more. It states that, *by definition*, h is the ratio of q to ΔT.

Eq. (4-3) should be frequently repeated throughout every text on the old heat transfer in order to remind the reader that h has no fundamental character, and that it is nothing other than an alias for (q/ΔT).

THE GRAPHICAL ANALOG OF h

Motion deals with distance (s) and time (t), and all motion analysis starts with

$$s = f\{t\} \tag{4-4}$$

In motion analysis, velocity is an important concept. Every good high school physics book promotes a real understanding of velocity by discussing its graphical analog. It is only the slightest exaggeration to say that everyone involved in science or engineering knows that the graphical analog of velocity at point A in Fig. (4-1) is the slope of the tangent at point A.

Convective heat transfer deals with heat flux (q) and boundary layer temperature difference (ΔT). If h has a graphical analog, it must be on a graph of q vs ΔT. The y and x coordinates in Figure (4-2) are in fact q and ΔT.

When asked to draw the graphical analog of h at Point B in Figure (4-2), the graphical answer most often given is:

> The graphical analog of h at Point B is
> the slope of the tangent at that point.

Is this answer correct? Is the graphical analog of h at Point B really the slope of the tangent at that point?

In Fig. (4-2), the slope of the tangent at Point B is the first derivative of q with respect to ΔT. If this derivative were in fact equal to h, the above answer would be correct. In other words, *IF* h were in fact defined by Eq. (4-5), then the above answer would be correct.

$$h = dq/d\Delta T \tag{4-5}$$

But of course h is not defined by Eq. (4-5), and therefore the above answer is *not* correct. What is h? How is h defined? We noted above that h is an alias for the dimensional group ($q/\Delta T$). Therefore, the graphical analog of h at a point must equal ($q/\Delta T$) at that point. And since the slope of a straight line drawn between (0,0) and any point in Fig (4-2) equals ($q/\Delta T$), the correct answer is:

> The graphical analog of h at Point B in Figure (4-2) is the slope of a straight line drawn between Point B and (0,0).

It is important to note that the slope of such a line has no use in pure mathematics. The information it contains is virtually useless--useless because it is information about combined rather than separated variables--useless because it is information about the ratio y/x. This slope which has *no use* in pure mathematics is the *foundation* of the old convective heat transfer!

h--A VARIABLE

In the old heat transfer, our thinking is focused on the equation "$q=h\Delta T$". This form strongly suggests that h is a constant in the equation because that is the convention we learned in pure mathematics.

When we write "y=mx", we mean that m is a constant in the equation--we mean that m is independent of x and y. If m is a function of x, we do not write "y=mx". We write simply "y=f{x}". We *intend* the equation "y=mx" to be a description of the functionality between y and x, and to indicate that m is a constant in the equation.

Because of the way we write equations in pure mathematics, it is natural to think of h as a constant in the equation "$q=h\Delta T$". But h is *not* a constant in this equation--h often depends on ΔT. Therefore we can *not* treat h as a constant in the equation. We must treat h as a variable. And therefore the correct answer to the question "What is $dq/d\Delta T$?" is:

$$dq/d\Delta T = h + \Delta T(dh/d\Delta T) \qquad (4\text{-}6)$$

When it is recognized that h is an alias for $q/\Delta T$, it becomes obvious that Eqs. (4-1), (4-2), and (4-6) are mathematically undesirable, and that their mathematical analogs play no role in pure mathematics.

(The interested reader may want to prove that Eq. (4-6) is an identity. To do this, substitute $q/\Delta T$ for h and show that the right side of the equation reduces to $dq/d\Delta T$.)

SEPARATING THE VARIABLES

In pure mathematics, we learn to "separate the variables"--ie to arrange all the y's in one place and all the x's in another. The reason we separate variables is not because that is the only way problems can be solved. The reason we separate variables is because that generally makes problems *easier to solve*. If we had no interest in simplifying the solution of problems, we would not bother to separate variables.

Eq. (4-1) is a complicated form of a very simple equation. The reason it is complicated is that it is written in terms of variables which are not separated. Instead of asking for the simple form of Eq. (4-1) in Problem 3 of the quiz, the problem could have been stated as "Separate the variables in Eq. (4-1)."

I have stated the problem in this way many times, and I have not found one person who had any idea how to separate the variables in Eq. (4-1). The reason no one knew how to separate the variables was because no one could identify the real variables in the equation. And the reason no one could identify the real variables in the equation is because no one really knew what h is.

The simple form of Eq. (4-1) is readily determined by separating the variables in the equation. The secret in doing this is to know that h is an alias for $q/\Delta T$, k is an alias for $(q\ \Delta x/\Delta T)$, and U is an alias for the ratio of q to total temperature difference. When we have made these substitutions, it becomes apparent that the real variables in Eq. (4-1) are q and ΔT. And when these variables are separated, it becomes apparent that the simple form of Eq. (4-1) is Eq. (4-7)!

$$\Delta T_{total} = \Delta T_1 + \Delta T_w + \Delta T_2 \tag{4-7}$$

It should be noted that Eq. (4-1) is an absurd way to write Eq. (4-7). The absurdity can be seen only if it is consciously recognized that h is an alias for $q/\Delta T$, k is an alias for $q\ \Delta x/\Delta T$, etc.

In the new heat transfer, we are consciously aware that h is an alias for $q/\Delta T$, and this awareness allows us to see that Eq. (4-1) is mathematically unsound. Therefore, we reject Eq. (4-1) and replace it with Eq. (4-7) which is mathematically sound.

It is important to note that *EVERY* problem that can be solved using Eq. (4-1) and the old heat transfer can also be solved using Eq. (4-7) and the new heat transfer. If the problem involves only proportional behavior, both methods are simple and effective. But if the problem involves nonlinear behavior, the solution is *MUCH* simpler using Eq. (4-7) and the new heat transfer.

CONCLUSION

There is only one correct and complete answer to the question "What exactly is h?"

h is exactly the ratio of q to ΔT.

(This completes the solution of the quiz on page 55. If you correctly answered more than one of the four questions, you did exceptionally well. Now turn to page 54 and answer the final question.)

CHAPTER 5

WHAT'S WRONG WITH h

Whenever h is written/spoken, what is actually being written/spoken is "the ratio of heat flux to temperature difference". And when you are consciously aware that h is nothing other than an alias for "the ratio of heat flux to temperature difference", the following questions arise:

- Why are we primarily concerned with the ratio $q/\Delta T$ (alias h) in heat transfer?

- We have no use for the ratio y/x in mathematics. Why use the ratio $q/\Delta T$ (alias h) in heat transfer?

- Why do we *combine* q and ΔT in heat transfer coefficients when we make every effort to *separate* y and x in mathematics?

- Why do we *generate* the ratio $q/\Delta T$ (alias h) in heat transfer when we make every effort to *eliminate* the ratio y/x in mathematics?

In this chapter, we consider these questions in order to answer the question which really concerns us:

What's wrong with h?

THE MATHEMATICAL ANALOG OF TODAY'S OLD HEAT TRANSFER

Texts on today's old heat transfer usually describe the electrical analog of heat transfer and note that (1/R) is the electrical analog of h. As we shall see in a later chapter, this analogy is misleading because today's h is NOT truly analogous to (1/R).

The h conceived by Fourier was analogous to (1/R). But when h ceased to be the proportionality constant between q and ΔT, and became the variable ratio ($q/\Delta T$), the analogy was largely destroyed.

There is a mathematical analog of the old heat transfer, but it is not found in any text on the old heat transfer. To establish the analog, we must relate the parameters q, ΔT, and h to the parameters y and x. From the viewpoint of the old heat transfer, this is an impossible task. How can three parameters be analogous to two parameters?

But from the viewpoint of the new heat transfer, it is a simple task to relate the three parameters q, ΔT, and h to the two parameters y and x. We are consciously aware that there are *two* parameters in q, ΔT, and h because h is merely an alias for $q/\Delta T$. Therefore the mathematical analog of the old heat transfer is given by:

OLD HEAT TRANSFER	MATH ANALOG
ΔT	x
q	y
h	y/x
$q = h\ \Delta T$	$y = (y/x)x$

The important thing to note is that the mathematical analog of h is (y/x)!

- What is (y/x) called in mathematics?

  Nothing! (y/x) has no name.

- Why doesn't (y/x) have a name?

  Because (y/x) is not a useful parameter.

- What is (y/x) used for in mathematics?

  Nothing.

- When do we plot (y/x) vs x?

  Never. We plot y vs x.

- When do we write y = (y/x)x?

  Never. We write y = f{x}.

- If we are given values of y and x, do we immediately divide y by x in order to generate (y/x)?

  No. We keep y and x SEPARATE!

- Why do we eliminate (y/x) in mathematics?

  Because (y/x) complicates the solution of problems by making it impossible to separate y and x.

- Since h is exactly like (y/x), what do the above questions say about h?

  They say that h is mathematically unsound.

If h is mathematically unsound--*and it is*--then we should abandon/eliminate/outlaw h.

And that is exactly what we do in the new heat transfer. We *abandon* h. We abandon an unsound mathematical foundation, and we replace it with a sound one. The end result is that we greatly simplify/improve the solution of heat transfer design and analysis problems.

THE MOTION ANALOG OF THE OLD HEAT TRANSFER

The motion analog of the old heat transfer is even more revealing than the mathematical analog because it is less abstract. In the old heat transfer, we have the parameters q, ΔT, and h (an alias for $q/\Delta T$). In motion analysis, we have s and t. Therefore the motion analog is:

OLD HEAT TRANSFER	MOTION ANALOG
ΔT	t
q	s
h	s/t
$q = h\,\Delta T$	$s = (s/t)t$

Notice that the motion analog of h is (s/t).

- What is (s/t) called in motion analysis?

  Average velocity.

- What is (s/t) used for in motion analysis?

  Very little.

- When do we plot (s/t) vs t?

  Never. We plot s vs t.

- When do we write s = (s/t)t?

  Never. We write s = f{t}.

- Do we embrace (s/t) in motion analysis?

  No. We avoid s/t because it prevents the separation of the variables s and t.

- Since h is *exactly* like (s/t), what do the above questions say about h?

  They say that h is mathematically unsound.

MOTION COEFFICIENTS

Let us extend the motion analog by formulating a "motion science" which is closely analogous to the old heat transfer. This new motion science is based on "Newton's Law of Motion":

$$s = mt \tag{5-1}$$

where s is distance traveled over elapsed time t, and m is the "motion coefficient". Eq. (5-1) is the defining equation for m. It is not a phenomenological description of motion because m often depends on t. Eq. (5-1) seems to state that s is proportional to t, but it actually makes no statement about s{t} functionality. It simply states that m is an alias for s/t.

Since m often depends on t, we must treat m as a variable in the equation. We must recognize that Eq. (5-1) contains *three variables*--s, m, and t. And this should come as a huge surprise because every reader knows that motion is completely described by *two* variables--s and t! In other words, when we use Eq. (5-1) to deal with motion in a general way, we must deal with three variables even though motion is actually described by only two variables!

- We have *invented* a variable--a parameter not found in Nature--the motion coefficient m.

- By inventing the motion coefficient m, we have transformed *every* motion problem from a problem in two variables into a problem in three variables!

- By inventing the motion coefficient m, we have complicated/confused the solution of *every* motion problem by making it necessary to think and analyze using three parameters even though motion problems actually contain only two parameters!

This new motion science and "Newton's Law of Motion" with its motion coefficient m are obviously undesirable. They are as undesirable as their analogs the old convective heat transfer science and "Newton's Law of Cooling" with its heat transfer coefficient h.

THE MOTION ANALOG OF THE NEW HEAT TRANSFER

In motion analysis, we deal with s and t. In the new convective heat transfer, we deal with q and ΔT. The motion analog of the new heat transfer is obviously:

NEW HEAT TRANSFER	MOTION ANALOG
ΔT	t
q	s
$q = f\{\Delta T\}$	$s = f\{t\}$

The motion analog of the new heat transfer is the motion science which has stood the test of time for centuries--the same mathematically rigorous science which makes it possible to aim at and hit the stars.

Twentieth century motion science is mathematically rigorous because it is founded on the Gibraltar of mathematics:

$$y = f\{x\} \tag{5-2}$$

Because the new heat transfer is founded on the Gibraltar of mathematics, it too is mathematically rigorous. And it will remain mathematically rigorous until Eq. (5-2) is abandoned for a new and better mathematical science.

CONCLUSION

This chapter has a single objective:

> To describe exactly what's wrong with h in the hope that the reader will convince himself that he should abandon today's old heat transfer.

The rest of this book is based on the assumption that this chapter satisfies its objective. We shall have little further discussion about the need to abandon/eliminate/ outlaw h.

CHAPTER 6

THE ORIGIN OF THE h CONCEPT

When I realized in 1963 that I had conceived a new heat transfer science which required that h be abandoned, I felt the need to answer the following questions:

- How was heat transfer behavior perceived by the person who had conceived h?

- How did the twentieth century come to have a heat transfer science which is based on a poor mathematical foundation?

NEWTON AND h

Heat transfer texts generally claim that Newton conceived the h concept, and cite his manuscript "A Scale of the Degrees of Heat" published in 1701.

The truth is that Newton's article had nothing to do with the heat transfer coefficient concept. Newton's alleged conception of h is mythical, and has no foundation in fact. It is historically inaccurate by over 100 years.

The reader can readily verify my singular contention that Newton did *not* conceive h:

- Look up Joseph Black in any large encyclopedia.

- Note that Black was the first to make a clear distinction between heat and temperature, and that he accomplished this in the year 1760.

- Note that h is the flow of *heat* per unit area per unit *temperature*, and therefore the h concept requires a clear distinction between heat and temperature.

- Consider the question: "How could Newton have conceived the h concept in 1701 when no clear distinction was made between heat and temperature until 1760--thirty-three years after Newton died?"

Newton's actual contribution to heat transfer science was his "Law of Cooling" which had nothing to do with heat transfer coefficients, and nothing to do with concepts which were new in his time.

The mathematical expression of the law of cooling described by Newton is:

$$\frac{dT}{dt} \propto -(T_{object} - T_{ambient}) \qquad (6\text{-}1)$$

Notice that Newton's law of cooling has nothing to do with heat or heat transfer coefficients. It contains only one thermal concept--temperature.

FOURIER AND h

Fourier actually conceived the h concept, and also the concepts of flux in general, heat flux in particular, the continuity of flux at boundaries, thermal conductivity, and dimensional analysis. It is the slightest exaggeration to say that the old heat transfer is Fourier's brainchild.

Herivel (1975) pays touching tribute to Fourier's genius:

> (The concept of heat flux) must be regarded as (Fourier's) most critically important and original insight into the physical nature of the conduction of heat in solid bodies. . . .

> Fourier's contemporaries (Laplace, Biot, and Poisson) found it excessively difficult either to understand or to accept this concept. (*And they refused to accept it for over a decade.*)
>
> . . . this is surely another example of one of those apparently simple, almost trivial, concepts in theoretical physics which nevertheless seem to require for their formulation the intervention of a Galileo or a Newton.

In his treatise on heat transfer, Fourier states that he conceived the concepts of heat transfer coefficient and thermal conductivity. Fourier was one of the greatest mathematicians of all time, and therefore it is not likely that the mathematical difficulty with today's h resulted because Fourier was lacking in mathematical skills. The difficulty must have arisen because of Fourier's perception of heat transfer behavior. Fourier describes his perception of heat transfer behavior in the following words:

> The differential equations of the movement of heat are mathematical consequences analogous to the general equations of equilibrium and of motion, and are derived like them from the most constant natural facts.
>
> The coefficients c, h, k, which enter into these equations, must be considered, in general, as variable magnitudes, which depend on the temperature or on the state of the body. But in the application to the natural problems which interest us most, we may assign to these coefficients values sensibly constant.

Fourier's perception of heat transfer behavior was that "the natural problems which interest us most" are described by heat transfer coefficients which are "sensibly constant".

Fourier did not intend $q=h\Delta T$ to be a definition of h. He *meant* it to be a phenomenological description of heat transfer behavior at solid/fluid interfaces. He *meant* it to be a statement that heat flux is proportional to temperature difference across an interface. He *meant* h to be the constant of proportionality between heat flux and temperature difference. He *meant* k to be the constant of proportionality between q and dT/dx.

Some time after Fourier, it was recognized that heat flux was *not* always proportional to interface temperature difference. And if heat flux was not always proportional to temperature difference, how could h be a constant of proportionality?

This anomaly was resolved by transforming "Newton's law of cooling" and h. "Newton's law of cooling" ceased to be a phenomenological description of convective heat transfer behavior, and became merely the expression which defines h. And h ceased to be the constant of proportionality between q and ΔT, and became an alias for $(q/\Delta T)$.

> *When "Newton's law of cooling" and h were transformed in order to deal with nonproportional heat transfer behavior, the mathematical foundation of heat transfer science was transformed from rock to quicksand.*

In the new heat transfer, we do not again transform "Newton's law of cooling" and h--we *abandon* them. We replace them with the free form equation

$$q = f\{\Delta T\} \qquad (6\text{-}2)$$

and its free form concept of thermal behavior. The particular advantage of free form concepts is that they make it possible to solve problems by separating the variables, and this in turn makes it possible to deal effectively with *ALL* types of behavior including highly nonlinear.

> *In going from the old to the new heat transfer, we again transform the mathematical foundation of heat transfer science. But this time, the transformation is from quicksand to bedrock.*

OTHER ARTICLES ON THE ORIGIN OF h

Adiutori (1989) examines the question of the true origin of h and concludes that Fourier should be credited with both h and "Newton's law of cooling", and also with other key concepts found in both the new and the old heat transfers (such as the heat flux concept).

Bergles (1988) examines the same question and concludes that Newton is correctly credited with both h and "Newton's Law of Cooling".

The interested reader will want to reach his own conclusion after reading both articles.

CHAPTER 7

DIMENSIONAL CONSISTENCY AND DIMENSIONS

The validity of "dimensional consistency" is so obvious that no serious effort is generally made to prove that Nature behaves in a dimensionally consistent way. It is obvious that Btu/hr ft^2 can not "equal" degrees Fahrenheit. Obviously, the dimensions on both sides of an equation must be identical. To question dimensional consistency is to question common sense.

In this chapter, we recall that history teaches that common sense is not a reliable guide in science. For centuries, and to Galileo's regret, common sense dictated that the Earth was the center of the universe. For centuries, common sense dictated that the Earth was flat.

Since common sense is unreliable in science, we conclude that common sense is not a sufficient basis on which to establish the validity of dimensional consistency. We must have a demonstration--a proof that dimensional consistency is in fact an accurate description of the way in which Nature behaves. Since this proof must be beyond question, it must be experimental and not intellectual.

In this chapter, we critically examine dimensional consistency, and we conclude that it is *not* valid because Natural phenomena do not generally exhibit dimensional consistency. We conclude this because nonlinear phenomena are obviously not dimensionally consistent, and Natural phenomena are oftentimes nonlinear.

Dimensional consistency is an invention--a product of the intellect--a Law which was drawn up and passed several centuries ago when the world seemed a proportional place--a Law which is beyond question only because it falls under the veil of common sense.

Because Natural phenomena do not generally exhibit dimensional consistency, we reject it in the new heat transfer. This rejection results in numerous changes in the way we view dimensions, the way we read equations, and the way we quantify Natural parameters.

DIMENSIONAL CONSISTENCY AND PROPORTIONAL BEHAVIOR

Dimensional consistency is an important part of the methodology used to describe Natural phenomena in today's old heat transfer. It is a requirement born in the nineteenth century when the known world seemed to behave in a proportional way.

In the nineteenth century, numerous experiments showed that Natural phenomena generally occur in a proportional way. The data indicated that convective heat flux was proportional to temperature difference; conductive heat flux was proportional to temperature gradient; voltage drop was proportional to amps; force was proportional to acceleration; stress was proportional to strain. It did indeed seem that the world was a proportional place--the data proved it!

For example, a convective heat transfer experiment might have resulted in the proportional data set below:

HEAT FLUX Btu/hr ft^2	TEMPERATURE DIFFERENCE F
0	0
3	1
6	2
9	3
12	4

TABLE (7-1) PROPORTIONAL DATA SET

Given the data, we must answer the question:

What shorthand method best describes/summarizes the data set in Table (7-1)?

Notice that the data set indicates q is proportional to ΔT, and therefore the data set is described by

$$q \propto \Delta T \qquad (7\text{-}1)$$

However, there is a difficulty with Relation (7-1). It is a qualitative description of the data set, and we desire a description which is quantitative.

The quantitative form of Eq. (7-1) is of course

$$q = a\ \Delta T \qquad (7\text{-}2)$$

where "a" is the constant of proportionality between q and ΔT, and is equal to 3.00 in the above data set. But we are unhappy with Eq. (7-2) because, if "a" is a pure number, then the equation seems to state that heat flux equals temperature difference, and common sense tells us that this makes no sense. It makes no sense to equate heat flux and temperature difference--no sense to write equations which are dimensionally inconsistent--which have dimensions on one side which are not identical to the dimensions on the opposite side.

We therefore conclude that "a" is not a pure constant--"a" must have dimensions--it must have whatever dimensions will make Eq. (7-2) "dimensionally consistent"--it must have the same dimensions as the group $q/\Delta T$. And if "a" has dimensions, then it must be a parameter. And rather than use the generic "a" to denote this parameter, we christen it "heat transfer coefficient" and assign it the symbol "h".

And that is in fact how proportional data and dimensional consistency and common sense logically result in "h" and "Newton's Law of Cooling" which should in truth be called Fourier's Law of Convective Heat Transfer:

$$q = h\,\Delta T \qquad (7\text{-}3)$$

This Law is dimensionally consistent, but not because the data says so--not because Nature says so. It is dimensionally consistent because *WE* say so! *WE* have made this Law dimensionally consistent by the logical, intellectual transformation of a pure number into a dimensioned parameter--by the transformation of the pure number "a" into the dimensioned parameter h--by the invention of the parameter h!

Why does h have the dimensions heat flux per unit temperature difference? Simply because these are the dimensions which appeal to common sense--these are the dimensions which bring dimensional consistency to Eq. (7-3)--which make the dimensions on the left side identical to those on the right side.

In the above example, dimensional consistency and the reasoning which leads to Eq. (7-3) seem logical enough. But in fact, the methodology is seriously flawed. The flaws are not readily apparent because the example describes the application of dimensional consistency to a phenomenon which exhibits proportional behavior, and proportional equations have only one pure number which must be transformed to a parameter, christened, and assigned dimensions.

But when this methodology is applied to nonlinear behavior, as in the example below, the flaws become readily apparent. Nonlinear equations contain several pure numbers, and each must be transformed to a different parameter and assigned various dimensions in order to preserve "dimensional consistency". The end result is that "dimensional consistency" is obviously irrational when applied to nonlinear phenomena because it requires the invention of a multitude of "parameters" which have no physical significance.

Rather than discuss the methodological flaws here, let us postpone the discussion until we have applied dimensional consistency to a nonlinear phenomenon. For now, let us simply note that the requirement of dimensional consistency *seems* logical enough when dealing with proportional behavior.

DIMENSIONAL CONSISTENCY AND NONLINEAR BEHAVIOR

Now let us apply dimensional consistency to a moderately nonlinear set of data. Suppose that an experiment on free convection heat transfer has resulted in the data set below:

HEAT FLUX Btu/hr ft^2	TEMPERATURE DIFFERENCE F
0.00	0.00
3.00	1.00
7.54	2.00
12.93	3.00
18.96	4.00

TABLE 7-2 NONLINEAR DATA SET

Correlation of the data set indicates it is described by:

$$q \propto \Delta T^b \qquad (7\text{-}4)$$

$$q = a\ \Delta T^b \qquad (7\text{-}5)$$

$$q = 3.00\ \Delta T^{1.33} \qquad (7\text{-}6)$$

Notice that there are two pure numbers in Eq. (7-5)--"a" and "b". If we are to conclude that "a" is a parameter, then surely we must also conclude that "b" is a parameter. If we assign a name to "a", then surely we must also assign a name to "b".

What should "a" be called? It cannot be called "heat transfer coefficient"--we have already assigned that name to the ratio of q to ΔT. What should "b" be called? If/when we assign names, symbols, and dimensions to "a" and "b", how will we describe their physical significance when they obviously have none?

Eqs. (7-2) and (7-5) describe similar phenomena. We point to Eq. (7-2) and state that "a" is really h. Since Eqs. (7-2) and (7-5) are so similar, it should be possible to point to Eq. (7-5) and state that something in the equation is really h. But of course that cannot be done because neither "a" nor "b" in Eq. (7-5) is the h parameter.

There is a dimensional difficulty with Eq. (7-5). If it is to be dimensionally consistent, then the "parameter" which replaces "a" must be assigned whatever dimensions are contained in $(q/\Delta T^{1.33})$. In this example, the "a" parameter must have the dimensions Btu/hr ft^2 $F^{1.33}$. But this is not satisfactory--it is not seemly for a "parameter" to have dimensions with fractional exponents.

And what will we do if another experiment should indicate that the value of "b" is 1.42? If we do not adjust the dimensions of the "a" parameter, then the equation which results from the second experiment will be dimensionally inconsistent. But if "a" must have a floating set of dimensions, can it truly be regarded as a parameter?

The "b" parameter will obviously be dimensionless. But is a dimensionless "parameter" really a parameter?

How are these difficulties resolved in the old heat transfer? By stating that "a" in Eq. (7-2) is a parameter, but "a" in Eq. (7-5) is *not* a parameter. The "a" in Eq. (7-5) is a special kind of number--a "dimensional number", meaning it is a number which has dimensions.

THE METHODOLOGICAL FLAWS IN DIMENSIONAL CONSISTENCY

The invention of "dimensional numbers" resolves the difficulties, but is this a satisfactory solution or a clever evasion? What in the world is a "dimensional number"? Isn't the phrase "dimensional number" itself a contradiction? Can a number really have dimensions and still be just a number?

A "dimensional number" is like a round square--there is no such thing. "Round" and "square" are mutually exclusive, and so are "dimensional" and "number". Dimensional numbers are a contradiction which makes dimensional consistency seem logical even though it is not. Dimensional numbers are an invention which excuses dimensional consistency from having to deal with every form of Natural behavior except strictly proportional.

How can we reason that "a" in Eq. (7-2) is a parameter and then conclude that "a" in Eq. (7-5) is *not* a parameter? Equations (7-2) and (7-5) are so similar that surely "a" possesses the same character in both equations. Either "a" is a parameter in both equations, or it is a dimensional number in both equations, or it is a pure number in both equations.

If dimensional consistency were truly valid, it would be able to cope with all types of Natural behavior. In fact, dimensional consistency can cope with only one type of behavior--the simplest of all possible behavior--proportional. If all Natural phenomena behaved in a proportional way, we might consider retaining dimensional consistency in the new heat transfer.

But Nature does *NOT* always exhibit proportional behavior. She often exhibits nonlinear behavior. And therefore methods which deal with Natural phenomena in a general way must be able to cope with nonlinear as well as proportional behavior. Dimensional consistency can *not* deal with nonlinear behavior. And for that reason, we reject it in the new heat transfer.

In the new heat transfer, we reject dimensional consistency. Therefore there is no requirement that "a" have dimensions. In the new heat transfer, "a" has the same character in both Eqs. (7-2) and (7-5), and that character is neither parameter nor dimensional number.

In both equations, "a" is just a number.

THE MEANING OF THE EQUAL SIGN IN THE NEW HEAT TRANSFER

In today's old heat transfer, equations and correlations must be dimensionally consistent in order to be regarded as scientific. But if we ask why this is so, the only answer is "because common sense tells us so". Common sense seems to verify the requirement of dimensional consistency because of the way we interpret the equal sign. The current rule is to interpret the equal sign to indicate numerical equality *AND* dimensional identity.

Now let us forget this old view and focus on the question which really concerns us:

> *What symbolism and symbol definitions will result in a correlation which makes exactly the same statement made by a data set?*

Let us answer this question by first formulating the statement made by a data set, then writing a correlation which quantitatively describes the data set, and finally defining the correlation symbols to obtain exact agreement between the statement made by the correlation and the statement made by the data set.

The data sets in Tables (7-1) and (7-2) are exactly described by the following statements:

> The heat flux in its specified dimensions is numerically equal to 3.00 times the temperature difference in its specified dimensions. (Table (7-1)).
>
> The heat flux in its specified dimensions is numerically equal to 3.00 times the temperature difference in its specified dimensions raised to the 1.33 power. (Table (7-2).)

The important thing to note is that, in both the data tables and in the above statements, the equality described is numerical equality and *NOT* dimensional identity. Data sets have *nothing to do* with dimensional identity. Therefore, any correlation which exactly agrees with a data set must make no statement about dimensional identity. Since we desire our correlations to exactly agree with data sets, they must have nothing to do with dimensional identity--nothing to do with "dimensional consistency".

Now look at Eq. (7-2). In order for Eq. (7-2) to exactly agree with the data set in Table (7-1), it must state:

> The heat flux in its specified dimensions is numerically equal to 3.00 times the temperature difference in its specified dimensions.

because this is exactly the statement made by the data set. In order to obtain this desirable result, we must redefine the meaning of the equal sign. The new meaning must be:

> *The equal sign means numerical equality.*
> *It makes NO statement about dimensional identity.*

This new definition of the equal sign may seem strange, but it results in correlations which *exactly* accord with data sets--exactly accord with Natural behavior--exactly accord with our stated desire.

When we obtained the data set, the experiment did not indicate that a heat flux of 6 Btu/hr ft^2 was *equal to* a temperature difference of 2 F. The experiment indicated that a heat flux of 6 Btu/hr ft^2 *resulted from* a temperature difference of 2 degrees F. Experiments make no statement about dimensional consistency. Data have nothing to do with dimensional consistency. Therefore there is no good reason why correlations which describe data should be dimensionally consistent.

In the new heat transfer, the equal sign indicates numerical equality and *NOT* dimensional identity. Therefore the correlations of the new heat transfer make no statement about dimensional consistency. The correlations of the new heat transfer are consistent with Natural phenomena because *NEITHER* has anything to do with dimensional consistency.

HOW TO READ AND WRITE EQUATIONS IN THE NEW HEAT TRANSFER

In order to correctly read and write equations in the new heat transfer, it is necessary to know that

- Parameter symbols such as q, T, P indicate the numerical value of the parameter *AND* its dimensions. Therefore the supporting nomenclature must specify the dimensions assigned to each symbol.

- Symbols are used *only* for fundamental parameters--only for parameters which actually exist. There are *no* symbols for dimensional or dimensionless groups--no symbols for contrived parameters. If a group of parameters is important, it is identified by the fundamental parameters which make up the group.

  For example, the old "Re" is replaced by xG/μ. The correlations of the new heat transfer contain only fundamental parameters because contrived parameters are not permitted--ie group parameters are not permitted.

- The symbol "=" indicates numerical equality and makes no statement about dimensional identity.

For example, if the nomenclature specifies that "q is heat flux in Btu/hr ft^2", and "W is flow rate in pps", then the equation

$$q = 7.23\ W^{.8} \qquad (7\text{-}7)$$

reads as follows:

> *The heat flux in Btu/hrft² is numerically equal to 7.23 times the flow rate in pps raised to the 0.8 power.*

In summary, the new way of reading and writing equations differs from the old way in that:

- New way correlations are not dimensionally consistent.

- New way symbols denote parameters *and* dimensions.

- New way equal signs indicate numerical equality. They make no statement about dimensional identity.

• New way correlations contain only fundamental parameters. Contrived parameters are not permitted.

The new way more closely accords with Natural phenomena, since neither has anything to do with dimensional consistency.

HOW TO QUANTIFY NATURAL PARAMETERS IN THE NEW HEAT TRANSFER

In the new heat transfer, symbols denote both the parameter and its dimensions. For example, we do not quantify heat flux by writing

$$q = 240 \text{ Btu/hr ft}^2$$

Since the dimensions are implicit in q, it would be redundant to state the dimensions explicitly. We write simply

$$q = 240$$

and the nomenclature states that the dimensions implicit in q are Btu/hrft^2.

In the same way, we write simply "T = 132" and "x = 2.7", and the nomenclature defines the dimensions implicit in T and x. This new system is used to quantify parameters in the remainder of this book.

CHAPTER 8

THERMAL BEHAVIOR--CONCEPT AND CORRELATION

Thermal behavior is the basis for all thinking/correlating/designing/analyzing in the new heat transfer. And although thermal behavior is a new concept, it requires no new parameters. It simply uses q and T in a new way.

- Thermal behavior is a description of heat transfer behavior which uses only two thermal concepts--q and T.
- Thermal behavior is a description of heat transfer behavior which has *nothing to do* with h, k, or ϵ.
- Thermal behavior is the generic function q{TDF} used to describe all types of heat transfer behavior.
- Thermal behavior is the function q{ΔT} used to describe convective heat transfer.
- Thermal behavior is the function q{dT/dx} used to describe conductive heat transfer.
- Thermal behavior is the function q{T} used to describe radiative heat transfer.

The thermal parameters q and T are the primary, dynamic parameters in the new heat transfer, and therefore they deserve/demand/receive our primary attention. But of course, we are also concerned with other parameters which influence the q{TDF} behavior which is our main concern.

It will be some time before thermal behavior correlations are generated within the framework of the new heat transfer. However, thermal behavior correlations are readily generated by translating the heat transfer correlations used in today's old heat transfer.

SEPARABILITY AND CONVECTIVE THERMAL BEHAVIOR

The experimental results in the literature on the old heat transfer implicitly warrant the following experimental conclusion:

> In convective heat transfer, the effects of fluid properties, fluid flow rate, geometry, and boundary layer temperature difference are essentially *separable*.

To illustrate, we may rewrite Eq. (8-1) in the form of (8-2) *provided* we somehow know or are willing to assume that the effects of x, y, and z on w are separable.

$$w = f\{x, y, z\} \tag{8-1}$$

$$w = f_1\{x\}\ f_2\{y\}\ f_3\{z\} \tag{8-2}$$

Eq. (8-2) is much simpler than (8-1), but we have no inherent right to substitute Eq. (8-2) because separability does not apply to all problems. Similarly, separability may not apply to the Natural phenomenon of interest. It must be demonstrated by experiment.

Since there is no good reason to doubt the experimental conclusion of separability suggested by the old heat transfer results, and since the task of determining thermal behavior correlations is greatly simplified when separability applies, we accept this experimental conclusion--*for now*. But we intend to verify separability when correlations are generated within the framework of the new heat transfer.

CONVECTIVE THERMAL BEHAVIOR

Of the three types of heat transfer phenomena, convective heat transfer is by far the most varied and complex. It includes one-phase and multi-phase fluids, forced and free convection, boiling and condensing. The correlations which describe it are similarly varied and complex.

Convective thermal behavior varies from essentially proportional (as in one-phase forced convection) to moderately nonlinear (as in one-phase free convection) to highly nonlinear (as in boiling).

When separability applies, thermal behavior correlations for one-phase convection *to a specific fluid* are of the form

$$q = f\{flow\}\ f\{T\}\ f\{geometry\}\ f\{\Delta T\} \qquad (8\text{-}3)$$

where the f's refer to unspecified functions, flow refers to fluid flow rate, and T refers to the bulk temperature of the fluid.

Correlations which apply only to one specific fluid are referred to as "fluid specific correlations", as opposed to "fluid general correlations". Fluid specific correlations are seldom used in the old heat transfer, but they are used almost exclusively in the new heat transfer.

The results reported in the literature on the old heat transfer generally indicate that $f\{\Delta T\}$ in Eq. (8-3) is given by:

$$f\{\Delta T\} = \Delta T \qquad \text{(forced convection)} \qquad (8\text{-}4)$$

$$f\{\Delta T\} = \Delta T^{1.33} \qquad \text{(free convection)} \qquad (8\text{-}5)$$

When separability applies, *fluid general* thermal behavior correlations for one-phase convection are of the form

$$q = f\{flow\}\ f\{prop\{T\}\}\ f\{geometry\}\ f\{\Delta T\} \qquad (8\text{-}6)$$

where prop refers to fluid properties evaluated at the bulk temperature, and the $f\{\Delta T\}$ function is described by Eqs. (8-4) and (8-5).

(The correlations of the old heat transfer usually evaluate fluid properties at the "film" temperature--ie at the average of the bulk and surface temperatures. In the new heat transfer, we avoid the use of the film temperature unless the data *clearly* requires it. In the old heat transfer, the use of film temperature is generally the result of assumption rather than demonstration.)

For example, Eqs. (8-3) and (8-6) might be in the form of Eqs. (8-7) and (8-8). Eq. (8-7) is fluid specific--it applies only to a particular fluid. Eq. (8-8) is fluid general--it applies to many fluids.

$$q = 140\ V^{.8}(1 + .012T_b)(1/D^{.2})\Delta T \qquad (8\text{-}7)$$

$$q = .023\ G^{.8}a_0^{.6}(C/\mu)^{.4}(1/D^{.2})\Delta T \qquad (8\text{-}8)$$

In Eq. (8-8), C and a_0 are defined by

$$C \equiv Q/W\Delta T \quad \text{as } \Delta T \to 0 \qquad (8\text{-}9)$$

$$a_0 \equiv q\ \Delta x/\Delta T \quad \text{as } q \to 0 \qquad (8\text{-}10)$$

where Q is heat flow rate in Btu/hr and W is fluid flow rate in #/hr. C and a_0 are pure numbers whose values depend in part on the dimensions of the parameters used in their determination. This would be of some concern in the old heat transfer because fluid general correlations are usually used. It is a matter of little concern in the new heat transfer because fluid general correlations are seldom used.

Eq. (8-7) is a "dimensional" equation, whereas Eq. (8-8) is "dimensionless". In the old heat transfer, there is a marked preference for dimensionless equations. But in the new heat transfer, both forms are acceptable. Our only concern is that correlations closely agree with the phenomena they purport to describe.

CONDUCTIVE THERMAL BEHAVIOR

In today's old heat transfer, conductive heat transfer is described by the equation:

$$q = -k\ dT/dx \qquad (8\text{-}11)$$

Contrary to "Newton's Law of Cooling" which does not describe behavior, today's view of Eq. (8-11) is that it *DOES* describe behavior, and in a *universal* way.

Now let us agree that Eq. (8-11) accurately describes the thermal behavior of all materials known today. What about tomorrow? Will tomorrow bring new materials which behave in a nonlinear way? Will today's materials behave in a nonlinear way at tomorrow's heat fluxes? Who is to say what tomorrow will bring?

Fourier did not foresee nonlinear heat transfer behavior. Ohm did not foresee nonlinear electrical behavior. Fourier and Ohm and their contemporaries considered "$q=h\Delta T$" and "$V=IR$" to be *universal and permanent Laws* which would accurately describe Natural phenomena *forever*. Newton's *Law* of Cooling. Ohm's *Law*. Not equations which describe some Natural phenomena some of the time, but *Laws* which describe *ALL* Natural phenomena *ALL* of the time, and *for* all time.

> *In the new engineering, we repeal these Laws and others like them because they do NOT universally describe Natural phenomena. Not today. Not tomorrow.*

With regard to k and Fourier's *Law*, the question arises: Should we retain the k concept because Eq. (8-11) *today* seems to be a universal phenomenological description of conductive heat transfer? Should we hold on to k until it becomes absolutely necessary to abandon it? Should we continue to use k until non-proportional materials have actually been developed and tested? Or should we abandon k *now* because the future may hold materials which will require us to abandon it?

And of course the answer in the new heat transfer is that we should *abandon k NOW*. And we do. We *abandon* k and we *repeal* Fourier's Law. We replace both of these proportional concepts with the free form concept for conductive thermal behavior in the new heat transfer:

$$q = f\{dT/dx\} \tag{8-12}$$

We point out that, for all known materials, f{dT/dx} is a proportional function. In other words, we state that all materials we know of today are described by

$$q = a(dT/dx) \tag{8-13}$$

where "a" is a constant--a pure number which has neither physical significance nor dimensions. We admit that we do not know whether Eq. (8-13) will always describe the heat transfer behavior of all materials. And we note that the discovery of materials which are not described by Eq. (8-13) will not discomfit us in the least because the generic expression, Eq. (8-12), encompasses *all forms* of material heat transfer behavior.

HOW TO TRANSLATE k INTO THE NEW HEAT TRANSFER

In order to translate k into the language of the new heat transfer, we must know exactly what k is and how its value is determined. Expressions such as "k = 7.3 Btu/hr ft F" mean that an experiment was performed to measure k, and this expression summarizes the data. In order to determine exactly what k is, we must know how thermal conductivity experiments are conducted and how the data are reduced.

In the old heat transfer, thermal conductivity measurements are based on the *assumption* that q and dT/dx are in fact proportional. The measurements are *not* made in a manner which would test or verify proportionality--ie q and dT/dx are not systematically varied in order to obtain the data required to appraise functionality.

Rather, data are usually obtained at one or several small values of q and dT/dx, and the data are reduced to the value of "k" by assuming proportionality. Because proportionality is assumed, the proportionality constant "k" is determined by dividing the measured value of q by the measured value of dT/dx. In short:

> k values are experimentally determined by measuring $q\Delta x/\Delta T$ *at small values of q*. Symbolically, k is obtained from
>
> $$k = (q\ \Delta x/\Delta T)_0 \tag{8-14}$$

where the subscript "0" indicates that data are obtained at small values of q. And the point is that Eq. (8-14) exactly describes k because it exactly describes how k values are generated.

Since Eq. (8-14) exactly describes k, it is the starting point for translating k. We would like to rewrite Eq. (8-14) using no thermal concepts other than q and T. This is accomplished by rewriting Eq. (8-14) in the form

$$a_0 = q\ \Delta x/\Delta T \tag{8-15}$$

where a_0 is a *constant--a pure number*--which is determined experimentally at small values of q. Eq. (8-15) has nothing to do with dimensional consistency, and a_0 is neither a dimensional number nor a parameter. It is a pure number whose value depends in part on the dimensions of q, T, and x. We read Eq. (8-15) in the following way:

> The pure constant a_0 equals the numerical value of q in its specified dimensions times $\Delta x/\Delta T$ in its specified dimensions. The 0 subscript indicates that "a" is measured at small values of q.

In the new heat transfer, we translate the old expression "k = 7.3 Btu/hr ft F" to the new expression

$$q = 7.3\ dT/dx \approx 7.3\ \Delta T/\Delta x \tag{8-16}$$

where q, T, and x are parameters which have physical meaning and specified dimensions, and 7.3 is a *pure number* without physical meaning or dimensions. We point out that Eq. (8-16) is based on data obtained only at small values of q, and therefore it applies at small values of q by demonstration, and at large values of q by assumption.

(Notice that, even if q were *not* proportional to dT/dx, there would be no ambiguity in Eq. (8-15) because it specifies that a_0 is evaluated *at very small values of q* (ie approaching zero). If the value of q were not specified, and if q were not proportional to dT/dx, ambiguity would result because the value of "a" would depend in part on the value of q at which it was evaluated.)

A LAST LOOK AT k

In the old heat transfer, the pure number 7.3 in the example above would be the parameter called "thermal conductivity" (k), and k would have physical meaning and dimensions. This old way of dealing with Natural phenomena has been widely used for 200 years, and therefore it seems quite perfect. But perhaps this old way will seem less than perfect if we ask and attempt to answer the question:

> How would we describe the conductive behavior of a material which did not behave in a proportional way? For example, how would we deal with conductive behavior described by Eq. (8-17)?

$$q = 8.3\ (dT/dx)^{1.50} \qquad (8\text{-}17)$$

In Eq. (8-17), thermal conductivity is *not* equal to 8.3. In order to force Eq. (8-17) into the thermal conductivity concept, we would have to state that Eq. (8-11) "is not a phenomenological description of conductive heat transfer behavior--it is a definition of k". In other words, we would have to redefine k so that it would no longer be the proportionality constant between q and dT/dx, but rather would become an alias for "the ratio of q to dT/dx". The end result would be that, for the material described by Eq. (8-17), k would be given by:

$$k = 4.10\ q^{.33} \qquad (8\text{-}18)$$

Eq. (8-18) should look absurd because it is absurd. It is a mathematically absurd way to deal with Eq. (8-17). It is in fact the coefficient way of dealing with nonlinear conductive thermal behavior.

We are coping here with the same question about k that we already answered about h: "If convective (conductive) thermal behavior is not universally proportional, should we retain the proportional concept h (k) and the proportional equation called "Newton's Law of Cooling" (Fourier's Law of Conduction)? Or should we abandon h (k) in favor of the free form concept inherent in $q=f\{\Delta T\}$ $(q=f\{dT/dx\})$?"

For the same impelling reasons which caused us to abandon h and to repeal "Newton's Law of Cooling", we also abandon k and we repeal "Fourier's Law of Conduction". We replace these proportional concepts and equations with the free form concept "thermal behavior" inherent in the free form equation of the new heat transfer, $q = f\{TDF\}$.

RADIATIVE THERMAL BEHAVIOR

As with conductive thermal behavior, the intent of the generic expression for radiative thermal behavior is that it be general enough to include all present and future forms of radiative behavior. The expression

$$q_{radn} = f\{T\} \qquad (8\text{-}19)$$

satisfies this requirement and is in fact the generic expression for radiative thermal behavior in the new heat transfer. It states that the science of radiative heat transfer is founded on the observation that radiative heat flux from the surface of a body is an unspecified function of the surface temperature of the body. We state that this functionality is influenced by the nature of the body surface, and that oftentimes but not always, the functionality is described by the simple function:

$$q_{radn} = a\,T^4 \qquad (8\text{-}20)$$

where "a" is a pure number and has no dimensions.

We stress that, although radiative thermal behavior is oftentimes more complex than Eq. (8-20), the generic expression, Eq. (8-19), is sufficiently broad to encompass all possible forms of radiative thermal behavior.

In the old heat transfer, the T exponent is generally taken to be 4, even if some other value would better describe the radiative behavior of a particular object. In such a case, the mismatch between 4 and the better value is generally taken up in the emissivity (ϵ). The end result is that ϵ is the ratio of q to σT^4.

In other words, ϵ is an alias for $q/\sigma T^4$, just as h is an alias for $q/\Delta T$, just as k is an alias for the ratio of q to dT/dx. And for the same impelling reasons which caused us to abandon h and k, we also abandon ϵ in the new heat transfer.

CHAPTER 9

HOW TO TRANSFORM COEFFICIENT CORRELATIONS INTO BEHAVIOR CORRELATIONS

The experimental methodology used to generate convective thermal behavior correlations differs considerably from that used to generate heat transfer coefficient correlations. But there is no need to repeat the old experiments in order to obtain thermal behavior correlations. They are readily obtained by transforming the heat transfer coefficient correlations of the old heat transfer as follows:

- Replace *all* dimensionless groups with fundamental parameters. For example, replace Re with xG/μ.

- Replace h with $q/\Delta T$. This is necessary in order to separate q and ΔT.

- Replace k with a_0. (This is only a matter of form, since k and a_0 are numerically equal.)

- Separate q and ΔT--ie rearrange the correlation so that q and ΔT are on opposite sides of the correlation.

- Rearrange the correlation into the form of Eq. (8-3) if a fluid specific correlation is desired, or into the form of Eq. (8-6) if a fluid general correlation is desired.

THE LANGUAGE OF BEHAVIOR

It should be stressed that tranformation from heat transfer correlation to thermal behavior correlation has no effect on the functionality described by the correlation, or on the accuracy of the correlation. Transformation changes form, not substance.

The new form reveals the functionality among fundamental parameters--the parameters of real concern to designers and analysts. It answers questions such as: How does the heat flux respond to the ΔT? How does the flow rate respond to the ΔP?

The old form reveals functionality among contrived parameters--the parameters of *NO* concern to designers and analysts. It answers questions such as: How is the Nusselt number related to the Reynolds number? How is the Stanton number related to the Prandtl number?

In the new heat transfer, we have no interest in these questions or their answers because we have no interest in the Nusselt number or the Stanton number or any other contrived parameter. Contrived parameters are not real--they are imaginary. We have no interest in imaginary parameters because the hardware we design and analyze is not imaginary.

We design and analyze *real* hardware, and therefore we care about *real* parameters. We do not make Nusselt number exchangers--we make *heat* exchangers. We do not make Reynolds number systems--we make fluid *flow* systems.

We care about heat flux and flow rate and temperature. We want to know how these and other real parameters are related to each other. That is the functionality which we need to know in order to deal with engineering phenomena in a simple, effective way. That is the functionality which *should* concern us--and which *does* concern us in the new heat transfer.

HOW TO TRANSFORM A GENERALIZED h CORRELATION

Suppose we are given the generalized heat transfer coefficient correlation:

$$Nu = .023\ Re^{.8}\ Pr^{.4} \qquad (9\text{-}1)$$

and we want to transform it into a generalized thermal behavior correlation--ie into a "fluid general" correlation. The transformation is accomplished as follows:

- Substitute hx/k for Nu, xG/μ for Re, Cμ/k for Pr:

$$hx/k = .023\ (xG/\mu)^{.8}\ (C\mu/k)^{.4} \qquad (9\text{-}2)$$

- Substitute q/ΔT for h and a_0 for k:

$$qx/a_0\Delta T = .023(xG/\mu)^{.8}(C\mu/a_0)^{.4} \qquad (9\text{-}3)$$

- Separate q and ΔT--ie rearrange the equation so that q is on one side and ΔT is on the other:

$$q = .023(a_0/x)(xG/\mu)^{.8}(C\mu/a_0)^{.4}\Delta T \qquad (9\text{-}4)$$

- Rearrange the equation into the form:

$$q = .023(G)^{.8}(a_0)^{.6}(C/\mu)^{.4}(1/x^{.2})\Delta T \qquad (9\text{-}5)$$

Eq. (9-5) is the new heat transfer transformation of the old heat transfer expression, Eq. (9-1). Notice that Eq. (9-5) is "user friendly" because it contains several designer controlled parameters *in explicit form.* The only implicit parameter in Eq. (9-5) which concerns us is fluid temperature, and that is necessarily implicit because we are dealing with a generalized correlation.

(The example below deals with a correlation which is "fluid specific". In that type of correlation, fluid temperature also appears in explicit form. In the new heat transfer, fluid specific correlations largely replace the generalized correlations of the old heat transfer.)

FLUID SPECIFIC THERMAL BEHAVIOR CORRELATIONS

Heat transfer correlations may be grouped into two broad categories:

- *Fluid general correlations* which apply to a large class of fluids.

- *Fluid specific correlations* which apply only to one particular fluid.

Correlations which are fluid specific are easier to use and are likely to be more accurate because they are based on fewer assumptions.

More than 99% of the heat transferred in the world is transferred to air or water. Air and water are the aqua regia of convective heat transfer. Because of the overwhelming heat transfer importance of air and water, it is surprising that there is no widely accepted fluid specific correlation for air or for water.

The reason for this omission is that, in the old heat transfer, the emphasis is on *generalized* correlations such as Eq. (9-1). This emphasis is so great that, even in industries which have used only one heat transfer fluid for decades, heat transfer data are correlated in a generalized way, and generalized correlations are used for design and analysis.

In the new heat transfer, we recognize that fluid specific correlations are much better/simpler/more accurate for design and analysis. Generalized correlations are useful only when it is necessary to deal with seldom used fluids.

In the new heat transfer, fluid specific correlations are much more widely used than generalized correlations. We generate fluid specific correlations not only for water and air, but also for many other fluids. In time, the better correlations are widely accepted.

DATA CORRELATION

Convective heat transfer data can be correlated in two considerably different ways:

- The data can be correlated in a general way, resulting in correlations which apply to a large class of fluids.

- The data can be correlated in a way which is specific to the heat transfer fluid in the experiment.

In the old heat transfer, the first method is the method of choice, and the second method is seldom used. In the new heat transfer, the opposite is true--the second method is always used, and the first method is used infrequently and in a secondary way.

HOW TO GENERATE FLUID SPECIFIC CORRELATIONS BY TRANSFORMATION

Until fluid specific correlations for convective thermal behavior are generated within the framework of the new heat transfer, they can readily be obtained by transforming the generalized correlations of the old heat transfer:

- Translate the generalized coefficient correlation into a generalized thermal behavior correlation as described above, resulting in an expression similar to Eq. (9-5).

- Group the fluid property terms together to obtain the function f{prop{T}}. Tabulate and then plot f{prop {T}} vs temperature to obtain F{T}.

- If possible, determine a simple analytical expression which approximates the graphical expression of F{T}.

- Substitute the analytical or the graphical expression of F{T} for the fluid property terms in the correlation obtained in the above first step.

For example, Eq. (9-5) would take the fluid specific form

$$q = .023(G)^{.8}F\{T\}(1/x)^{.2}\Delta T \qquad (9\text{-}6)$$

For the fluid of interest, F{T} is obtained by plotting $a_0^{.6}(C/\mu)^{.4}$ vs T.

THE ADVANTAGES OF FLUID SPECIFIC CORRELATIONS

Equations (9-1) and (9-6) are identical--they describe exactly the same behavior. They differ only in form. Eq. (9-1) is in the old heat transfer form--a fluid general form written in the opaque language of contrived parameters. Eq. (9-6) is in the new heat transfer form--a fluid specific form written in the transparent language of fundamental parameters.

Notice that Eq. (9-6) readily reveals the q{ΔT} functionality and also the manner in which this functionality is affected by flow rate, temperature, and geometry. On the other hand, notice that Eq. (9-1) readily reveals *nothing about any* of the parameters which truly concern us.

Because Eq. (9-6) is written in the language of fundamental parameters, it readily reveals how designer controlled parameters influence the performance of the equipment--readily reveals how the equipment design can be optimized. Because Eq. (9-1) is written in the language of contrived parameters, it readily reveals nothing about designer controlled parameters, in spite of the fact that the sole purpose of such correlations is to aid in design/analysis.

The fluid specific form is also a better form for data correlation because it allows us to correlate the parameters we actually measure in heat transfer experiments, rather than parameters we do not measure. For example, generalized correlations necessarily include fluid properties, and fluid properties are seldom measured in heat transfer experiments.

SUMMARY

- Fluid general thermal behavior correlations are readily obtained by translating the h correlations of the old heat transfer.

- Fluid specific thermal behavior correlations are readily obtained by translating the h correlations of the old heat transfer.

- In the new heat transfer, fluid general thermal behavior correlations are seldom used. Fluid specific correlations are used almost exclusively.

- In the new heat transfer, heat transfer data are always correlated in a way which is fluid specific. The data are sometimes also correlated in a fluid general way.

CHAPTER 10

POWER LAWS, LOG LOG GRAPHS, AND REGIMES

In today's old heat transfer, engineering correlations are predominantly power laws--ie they are generally of the form

$$y = mx^n \qquad (10\text{-}1)$$

And since power laws are straight lines on log log graph paper, virtually all graphs in the texts and literature on the old heat transfer are log log graphs.

In this chapter, we consider the question:

> Does the pervasiveness of power laws and log log graphs in the old heat transfer reflect functionality expressed by the data? Or does it reflect an *assumption* about functionality?

Our analysis indicates that the pervasiveness of power laws and log log graphs in the old heat transfer is the result of assumption and *not* the demonstrated result of Nature's expression. The power law assumption *may* often be true, but its validity is seldom demonstrated in the old heat transfer.

In the old heat transfer, data are plotted on log log graphs as a matter of course, and then straight lines are drawn through the data or through some region of the data. But this does not demonstrate the accuracy of the power law assumption. It merely determines the power law which best agrees with the data--and overlooks the fact that the best power law may not be the best correlation for the data. And of course our real objective is to obtain the best correlation--not the best power law.

Since power laws are the result of *assumed* behavior, every power law correlation is suspect if the validity of this assumption has not been verified. Since the power law assumption is seldom verified in the old heat transfer, we will not be surprised to find power law correlations which do not agree with the underlying data.

In the later chapters, we examine several power law correlations which are widely used in the old heat transfer, and we find that they do *NOT* agree with the data they purport to describe. We conclude that *ALL* power law correlations of the old heat transfer must be critically examined in order to identify those which must be rejected because they do not agree with the data they purport to describe.

POWER LAWS

The correlations of the old heat transfer are predominantly power laws--ie equations which resemble Eq. (10-1). For example,

$$Nu = .023\ Re^{.8}\ Pr^{.4} \qquad (10\text{-}2)$$

Correlations such as Eq. (10-2) are generally the result of the following methodology:

- Determine the important dimensionless groups using a deductive process called "dimensional analysis". This process has nothing to do with data. It is generally performed as part of the preparation for experiment. The Nu, Re, and Pr dimensionless groups in Eq. (10-2) are products of dimensional analysis.

- Perform an experiment in which the values of the dimensionless groups are varied over a wide range. It is not necessary to vary all the fundamental parameters which influence the phenomenon. It is necessary only to vary the "important dimensionless groups", and this can be accomplished in whatever manner is easiest. For example, if it is a bother to vary Re by varying the flow rate, then the experiment is performed at a single value of flow rate, and Re is varied by varying x.

- Plot the data or results on log log paper.
- Draw a straight line or lines through the data.
- Determine the analytical expression of the straight line or lines. For example, Eq. (10-2) is a set of straight lines on a log log graph of Nu vs Re. Each line is the result of a different value of Pr.

This methodology may result in the power law correlation which best agrees with the data. But how do we know that a power law is the best correlation form? Why should we consider only power laws? Why not try to determine the best correlation in a manner which allows *all* correlation forms to be considered?

The best power law is the best correlation only if the phenomenon in question does in fact exhibit power law behavior. But we have no *a priori* way of knowing whether or not a particular phenomenon exhibits power law behavior.

If it could be demonstrated *in a general way* that Natural phenomena exhibit power law behavior, then the above methodology would be a reliable way to generate correlations which accurately describe the underlying data (assuming there were no bias in the data).

Unfortunately, it *cannot* be demonstrated in a general way that Natural phenomena exhibit power law behavior. The generality of power law behavior can be *assumed*, but it cannot be demonstrated. Therefore, in each particular case, the parametric functionality must be considered an open question to be decided by experiment. And we must be ready to accept whatever functionality is indicated by the data because data is the only reliable road to a knowledge of functionality.

The methodology of the old heat transfer does *NOT* treat parametric functionality as an open question, and therefore it is appropriate to ask the following questions about each power law correlation from the old heat transfer:

- Was power law behavior indicated by the data? Or was it merely assumed?

- Was the data plotted on log log paper because the data indicated power law behavior? Or was log log paper used merely because data are usually plotted on log log paper?

- Were straight lines drawn through the data points because they described straight lines? Or did the points describe curved lines which were merely approximated with straight lines?

- Were straight lines drawn through the complete range of the data? Or were multiple regimes arbitrarily inferred from pronounced curvature on log log paper? And was a straight line segment then drawn only in one regime--one narrow, arbitrarily selected region of the data?

It is reasonable to ask these questions, but it is seldom possible to answer them because the texts and literature on today's old heat transfer contain very little data. Data is relatively unimportant in the old heat transfer, as evidenced by the fact that the heat transfer literature contains virutally no data. This situation is reversed in the new heat transfer which recognizes the paramount importance of data.

In the later chapters, we search the old heat transfer literature for data on nucleate boiling, transition boiling, and film cooling. We find that the data which has made its way into the literature bears little resemblance to the power law correlations widely used to describe these phenomena. We therefore reject these widely used power law correlations, and we conclude that *ALL* power law correlations from the old heat transfer are highly suspect.

LOG LOG GRAPH PAPER--TYPE 1 ERROR

If log log graph paper is used simply to plot data without reaching any conclusion about functionality, no error in logic can be made--ie there can be no erroneous conclusion unless there is some conclusion.

But if log log paper is used to reach a conclusion about functionality based on straight lines drawn through data points, the vagaries of log log paper can and often do cause errors. To avoid these errors, it is essential to be consciously aware of the following:

- *ALL* straight lines on log log paper have the form of Eq. (10-1).

- Eq. (10-1) is a particular form of the general form

$$y = mx^n + b \qquad (10\text{-}3)$$

- Eq. (10-3) is a *curved* line on log log paper *unless* b is precisely 0.00.

- If straight lines are drawn through data which describes Eq. (10-3) with $b \neq 0.00$, the value of "n" indicated by the straight line will *NOT* equal the value of n described by the data.

- In summary, if the data describe a function which does not pass through (0,0) or $(0,\pm\infty)$, the old heat transfer methodology leads to values of n and b which do *NOT* agree with the data because *ALL* power laws pass through (0,0) or $(0,\pm\infty)$.

The error described above is not small. For example, for 40 or 50 years, it has been widely agreed that nucleate boiling data are highly correlated by

$$q = a\, \Delta T^n \qquad (10\text{-}4)$$

where n is "3 or 4". But as we shall see in a later chapter, the data in the literature are highly linear and the true value of n is actually 1.

There is no additive constant in Eq. (10-4), but there should be. The nucleate boiling data in the literature indicate that nucleate boiling correlations like Eq. (10-4) should contain a finite additive constant. This missing constant agrees with thermodynamic reasoning which concludes that boiling requires a *finite* ΔT, and therefore that the nucleate boiling function does *NOT* necessarily pass through (0,0).

In summary, there is strong *a priori* evidence that nucleate boiling is not described by a power law because "b" in Eq. (10-3) is finite. But in the old heat transfer, the conviction that Natural phenomena are described by power laws is so overwhelming that strong *a priori* evidence is disregarded, and a power law is used even for nucleate boiling.

LOG LOG GRAPH PAPER--TYPE 2 ERROR

The Type 1 error described above occurs when the curvature on log log paper is small enough to be obscured by the scatter in the data. Type 2 error occurs when the curvature is readily apparent.

In the old heat transfer, the assumption that Natural phenomena exhibit power law behavior is widely accepted with conviction. This conviction is so strong that, even when curvature is readily apparent on log log paper, there is a tendency to correlate with power laws. This seeming contradiction is resolved by attributing the curvature to a "change in regime", and concluding that a different power law is required for each regime.

The limits of the regime are determined *NOT* by some fundamental change observed in the process (such as the change from laminar to turbulent). Rather, the limits of the regime are arbitrarily selected so that an acceptably small error will result from drawing a straight line segment through the curving data points.

The end result is that the data is "correlated" only in a narrow region, using a power law correlation which does not at all resemble the data. And we excuse ourselves from dealing with the data outside this narrow region because it is not in the "regime" of interest.

The example used in a later chapter to illustrate this type of error is the phenomenon known as "film cooling". For 40 or 50 years, film cooling data have been widely correlated in the form

$$\eta = a(x/ms)^b \tag{10-5}$$

where b is negative. This correlation is used only for the "away from injection point regime" because η vs x/ms data plotted on log log paper indicate pronounced curvature.

If only the region far away from the injection point is correlated, the error which results from drawing a straight line through curved data points is made acceptably small by adjusting the lower limit of the regime. We select the maximum error we find acceptable, and then we define the lower limit of the regime to be that point at which the difference between curved data and straight line segment gives the maximum error we selected.

The casual invention of regimes in today's old heat transfer is an expedient which makes it possible to use power laws to "correlate" data, even though the data in no way resembles power law behavior. Using the methodology of the old heat transfer, a set of film cooling data seemed to support the conclusion that there are two regimes in film cooling, and that power laws correlate the data in one of the regimes.

Using the methodology of the new heat transfer, *the SAME data* results in the conclusion that there is only one film cooling regime because *all the data* are well correlated by *ONE* expression--Eq.(10-6)--and it is *NOT a power law*.

$$\eta = 1/(1 + ax/ms) \tag{10-6}$$

Eq. (10-6) little resembles the power law of Eq. (10-5). For example, $\eta\{x=0\}$ is infinity based on Eq. (10-5), and unity based on (10-6). There is *overwhelming* a priori evidence that film cooling is not described by power laws like Eq. (10-5) because it is *physically impossible* for η to be greater than unity. Yet the power law conviction is so deep in the old heat transfer that power laws are usually used even for film cooling, and the maximum error is controlled by stating that the power laws are not to be used below some arbitrarily selected lower limit of x/ms.

LOG LOG GRAPH PAPER--TYPE 3 ERROR

Type 3 error concerns the presence or absence of data. When a line is drawn through plotted data points, it is strongly implied that the function is continuous along the line, and that it would be possible to obtain the data at any point along the line. It is also strongly implied that there are sufficient data points along the line to establish that the line does in fact describe data points--ie the line is supposed to describe data rather than assumption.

Suppose we perform an experiment in which the levels of x are 10, 20, 35, 50, 99, 100, and we plot the data on linear paper. The absence of data in the interval 50 to 99 will be apparent because there will be no data in a region which is 54% of the total length of the line. Because of the obvious absence of data, we will doubt that the line accurately describes the behavior of the phenomenon, and we will suspect that the absence of data reflects an inability to obtain data in the interval $50 < x < 99$.

When the same data is plotted on log log paper, the absence of data is much less apparent because the interval from 50 to 99 represents only 29% of the total length of the line. The end result is that we may accept a correlation which in no way resembles the underlying data simply because log log paper makes it appear that the data is continuous when in fact it is not. In other words, the seeming agreement between data and correlation may actually reflect the fact that there is no data over most of the range tested!

The example used in a later chapter to illustrate the Type 3 error is taken from the literature on "transition boiling". In the example, three theses performed at a world class technical institute over several decades all resulted in the conclusion that:

> Transition boiling data are highly correlated by straight lines drawn on log log paper between the points of maximum and minimum heat flux.

This conclusion resulted because it was assumed that the data was continuous through the transition region, when in fact very little transition region data was obtained. The lack of data was in part obscured because the data was plotted on log log paper, and Type 3 error resulted.

HOW TO AVOID CORRELATION ERRORS CAUSED BY LOG LOG PAPER

The correlation errors cited above all result from drawing straight lines through data plotted on log log paper, and then supposing that the straight lines do in fact summarize and correlate the data. There is a very simple way to avoid these and other correlation errors which can and often do result from using log log paper to correlate data:

> *NEVER use log log paper to induce functionality from data. ALWAYS use linear paper, even if you are certain the phenomenon of interest exhibits power law behavior.*

Linear paper should be used even when dealing with power law phenomena because there may be bias in the data set, and the bias will generally make the data set resemble Eq. (10-3) rather than (10-1). If the data is plotted on linear paper, the bias will be readily apparent because the line drawn through the data points will not pass through the origin as would a power law of positive slope.

In the event the data is plotted on linear paper and it becomes evident that the data truly describes a power law, it will not be necessary to replot the data on log log paper. It is a simple matter to plot power laws on linear paper, particularly if one has access to computer graphics.

In summary, linear paper rather than log log paper should always be used to correlate data because the correlation errors which are difficult to avoid when using log log paper are almost impossible to commit when using linear paper.

CONCLUSIONS

The correlation methodology of the old heat transfer is based on the implicit and generally unverified assumption that Natural phenomena exhibit power law behavior, and this assumption underlies the pervasiveness of power law correlations and log log graphs. The old methodology is based on correlating data by drawing straight lines through data plotted on log log paper, a technique which can and often does lead to correlations which little resemble the data they purport to describe.

In the new heat transfer, we abandon the assumption of power law behavior, and we induce functionality from the data. We recognize that it is difficult if not impossible to "see" the functionality of data plotted on log log paper, and so we replace the log log paper of the old heat transfer with linear paper because it makes functionality transparent.

The end result is that we use many correlating functions including power laws in the new heat transfer. This new methodology greatly improves the likelihood that correlations will agree with the data they purport to describe.

CHAPTER 11

AN INTRODUCTION TO THERMAL STABILITY

Today's old heat transfer implicitly views heat transfer as a static process, and static processes are inherently "stable" by definition. Stability and instability are relevant only to dynamic processes, and therefore the old heat transfer logically has nothing to do with stability or instability.

Adiutori (1964a) analyzed the stability of the heat transfer process. That article was a page taken from the new heat transfer, and it was the first presentation and derivation of the generic criterion for the stability of the heat transfer process. It assigned the generic name "thermal stability" to the stability of the heat transfer process.

Kovalev (1968) presents the generic criterion and correctly demonstrates its application to a temperature controlled system. He refers to Kovalev (1964) which presumably also presents the generic criterion. Another independent effort is the stability work by Stephan. Stephan (1968) indicates that he derived and presented the generic criterion in Stephan (1965).

The end result is that, in the area of thermal stability, some "new heat transfer" has found its way into the current literature. But even in the area of thermal stability, the new heat transfer has not yet found its way into current texts on the old heat transfer--for example, Holman (1986), Kreith and Bohn(1986), and Incropera and Dewitt (1985).

In this chapter, we deal with thermal stability in a simple but rigorous way. We define stability in general and thermal stability in particular, and then we analyze idealized and real systems to determine criteria which enable us to appraise system thermal stability.

We also note that thermal instability results in either hysteresis or undamped oscillatory behavior. We discuss hysteresis in this chapter, but we postpone the discussion of undamped oscillatory behavior to a later chapter.

THE MEANING OF "STABILITY"

Let us forget heat transfer for the moment and concentrate on the meaning of "stability". In this chapter, we shall use the following working definition of stability:

> *At a potential steady-state operating condition, a system is "stable" with respect to a particular parameter at a particular location if the system inherently opposes a vanishingly small change (perturbation) in the parameter at that location, and moves toward the unperturbed condition.*

(This general definition actually defines the stability of "hard" systems. We do not discuss the stability of "soft" systems in this book because all systems are hard with respect to thermal stability.)

For example, suppose that, at a potential steady-state operating condition, the temperature at a particular location in a system is 152 F. If the temperature at that location is suddenly perturbed to 153 F, the question of stability depends on whether or not the system opposes this perturbation and moves toward 152 F.

If the resultant imbalance in the system causes the temperature to decrease from 153 F, the system is stable with respect to temperature at that location. If the resultant imbalance does not cause the temperature to decrease from 153 F, the system is unstable with respect to temperature at that location, and the temperature either remains at 153 F or continues to increase.

(Notice that we reject the conventional notion of "null stability" between stable and unstable--the condition in which perturbations neither increase nor decrease. In the new heat transfer, null stability is simply unstable.)

A FLUID FLOW ANALOG OF THERMAL STABILITY

In the new heat transfer, we view heat transfer as a dynamic process in which heat flows in response to a thermal driving force. Since the "flow" of heat is an abstraction, it is helpful to utilize a less abstract analog such as a fluid flow analog.

Suppose a water system includes a large, open vessel which acts as an in-line surge tank. Water continuously flows into and out of the tank as indicated in Figure (11-1). Further suppose that the level (L, inches) in the tank is 74 inches and that, at this level, the flow rate (W, pps) into and out of the tank are both 100 pps.

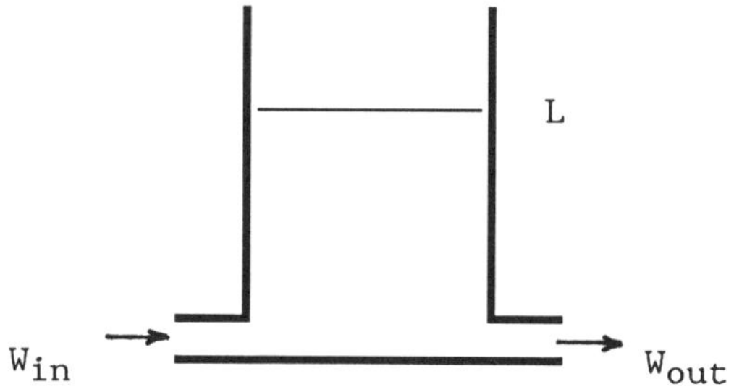

FIGURE (11-1) FLUID FLOW SYSTEM

Since the flow rate into the tank equals the flow rate out, this condition is a *potential* steady-state condition--ie it is a steady-state condition *if the system is stable at this condition.*

Now suppose a workman empties a bucket of water into the tank, causing the water level to increase to 74.1". The stability question is:

> Will the level move toward 74.0"? Or will the level remain at or increase above 74.1"?

In other words, will the system resist this perturbation and move toward the unperturbed condition in which the level is 74.0" and the flow rate is 100 pps? If so, we will conclude the system is stable with respect to level *at a level of 74"*. If not, we will conclude the system is unstable with respect to level at a level of 74".

We can answer this stability question if we know how the flow rate into the tank at 74.1" compares with the flow rate out of the tank at 74.1 inches. Symbolically, we wish to know whether Criterion (11-1) or (11-2) is satisfied:

$$W_{in}\{L=74.1\} < \text{s}\ W_{out}\{L=74.1\} \qquad (11\text{-}1)$$

$$W_{in}\{L=74.1\}\ \text{u}\geq W_{out}\{L=74.1\} \qquad (11\text{-}2)$$

If Criterion (11-1) is satisfied, the flow rate into the tank is less than the flow rate out, and thus the level decreases. This demonstrates that the system opposes small perturbations in L at L=74, and that the system is stable with respect to L at L=74. (<s is stable, <u is unstable.)

On the other hand, if Criterion (11-2) is satisfied, the flow rate into the tank is equal to or greater than flow rate out, and the tank level does *NOT* decrease toward the unperturbed value. Therefore the system is unstable with respect to L at L=74.

For example, suppose that

$$W_{in}\{L=74.1\} = 99.9$$

$$W_{out}\{L=74.1\} = 99.7$$

Therefore, after the workman throws the bucket of water into the tank,

$$W_{in} - W_{out} = (99.9 - 99.7) = .2 \qquad (11\text{-}3)$$

Since the storage rate in the tank is positive, the level in the tank will increase above 74.1 inches. This direction is *away from* the unperturbed condition, and thus we conclude the system is unstable with respect to L at L=74. We cannot simply state that the system is unstable with respect to L because the system might very well be stable at levels other than 74 inches. The given information is sufficient only to state that the system is unstable with respect to L at L=74.

Now suppose we do not know how W responds to L, but we know how ΔW responds to ΔL at L=74. Suppose we know that

$$(dW_{in}/dL \text{ at } L=74) = -.1$$

$$(dW_{out}/dL \text{ at } L=74) = -.3$$

We can appraise the stability in essentially the manner described above by noting that the criteria

dW_{in}/dL <s dW_{out}/dL (11-4)

dW_{in}/dL u≥ dW_{out}/dL (11-5)

establish the direction of movement following an incremental change in level. If Criterion (11-4) is satisfied at L=74, a positive change in L results in a negative storage rate and a decreasing water level, and we conclude that the system is stable with respect to L at L=74. On the other hand, if criterion (11-5) is satisfied at L=74, we conclude the system is unstable with respect to L at L=74.

THE EFFECTS OF INSTABILITY

If the system in the above example is stable, it will respond to a small perturbation in level by returning to the unperturbed condition.

But if the system is unstable, how will it respond to a perturbation in level? Notice that the above analysis does not address this question. In order to quantitatively describe the behavior of an unstable system, it would be necessary to have considerably more information than is given in the example above. It would be necessary to know how the flow into and out of the tank respond to level over a considerable range.

Later in this chapter, we discuss the behavioral analysis of systems which are unstable. But it should be noted here that instability results in either hysteresis or undamped oscillatory behavior. In the example problem above, instability would result in one of two alternatives, depending on the W{L} functions:

- The level in the tank would settle out at some value other than 74 inches. The final level would depend on the sign of the perturbation and on the W{L} functions.
- The level in the tank would never settle out, but would continue to oscillate about L=74. The oscillation amplitude would depend primarily on the W{L} functions.

THE BEHAVIOR SUGGESTED BY THE OLD ENGINEERING CONCEPTS

In the old engineering, the fundamental concepts are contained in equations such as

$$q = h \, \Delta T \qquad (11\text{-}6)$$

$$\Delta P = f \rho V^2 x / 2gD \qquad (11\text{-}7)$$

These equations *seem* to describe the behavior of flow vs driving force. Eq. (11-6) *seems* to state that q will increase if ΔT is increased. Eq. (11-7) *seems* to state that V will increase if ΔP is increased.

But the reality is that Eqs. (11-6) and (11-7) do *NOT* describe the behavior of flow vs driving force. They *make no statement* about the behavior of flow vs driving force.

Eq. (11-6) seems to describe q{ΔT}--it seems to state that q is proportional to ΔT. Yet it is widely agreed that, in one phase free convection, q is related to ΔT to the 4/3 power. In nucleate boiling, it is widely agreed that q is related to ΔT to the power of "3 or 4". And in transition boiling, it is widely agreed that q DEcreases with INcreasing ΔT--precisely the *opposite* of the behavior Eq. (11-6) seems to suggest.

Eq. (11-7) seems to describe ΔP{V}--it seems to state that ΔP is proportional to V squared, Yet we know from experiment that, in laminar flow, ΔP is *linearly* related to V. And in two phase flow, we know that the ΔP often DEcreases as flow INcreases--precisely the opposite of the behavior suggested by Eq. (11-7).

The explanation for these seeming anomalies is that Eqs. (11-6) and (11-7) *make no statement* about the behavior of flow vs driving force--they merely define h and f. Eq. (11-6) defines h to be the ratio of flow to driving force. Eq. (11-7) defines f to be proportional to the ratio of driving force to flow squared.

> *Eq. (11-6) only SEEMS to state that the flow is proportional to the driving force. It actually states that the flow is some unspecified function of the driving force.*

> *Eq. (11-7) only SEEMS to state that the driving force is proportional to the flow squared. It actually states that the driving force is some unspecified function of the flow.*

If our thinking is based on the flow vs driving force behavior these equations *SEEM* to describe, then our thinking will be wrong whenever we deal with phenomena which do not behave in the manner these equations seem to describe.

For example, I have heard engineers of considerable engineering education and experience express the opinion that laminar flow is generally undesirable because it requires more pumping energy to transport fluids in laminar flow. The rationale behind this conclusion is:

- The friction factor is very large in laminar flow.
- Pressure drop is proportional to friction factor.
- Pumping power is proportional to pressure drop.

It is quite true that the "friction factor" is very large in laminar flow. It is quite true that pumping power is proportional to pressure drop. But it is NOT true that more energy is required to pump fluids in laminar flow than in turbulent flow. A great deal LESS pumping energy is required if the flow is laminar.

Eq. (11-7) seems to state that ΔP is proportional to f, but it makes no such blanket statement. Since f is a function of V, Eq. (11-7) actually states that, *for a given velocity*, ΔP is proportional to f. But this is not germane to the discussion--it has nothing to do with the pumping power required for laminar vs turbulent flow.

There is a natural tendency to interpret Eq. (11-7) to mean that ΔP is proportional to f because that is the way we would interpret Eq. (11-7) in mathematics. But Eq. (11-7) does *NOT* state that ΔP is proportional to f--it states that

$$\Delta P \propto f\{V\}\ V^2 \qquad (11\text{-}7a)$$

because the friction factor is a function of V. Therefore Eq. (11-7) is an absurd way to write "ΔP is an unspecified function of V".

The absurdity results because *BOTH* ΔP and f depend on V. This is not apparent from Eq. (11-7) because there is no indication that f is a function of V. Eq. (11-7) is better written in the form of Eq. (11-7b) because this form better reveals the mathematical absurdity by explicitly stating that f is a function of V.

$$\Delta P = f\{V\}\rho V^2 x/2gD \qquad (11\text{-}7b)$$

The equation "$y=mx^2$" means that y and x are variables and m is a constant in the equation. It would be absurd for "$y=mx^2$" to mean "y and m are both variables which depend on x, and m is the variable defined by the ratio of y to x squared". Yet that is in fact the precise analog of Eq. (11-7) because f is not a constant in the equation--f is a variable which depends on V.

We mentally assign to Eq. (11-7) the sensible meaning it would have in mathematics rather than the absurd meaning it has in fluid flow. That is why Eq. (11-7) seems to describe $\Delta P\{V\}$ when it does not. Eq. (11-7) makes no statement about $\Delta P\{V\}$, just as Eq. (11-6) makes no statement about $q\{\Delta T\}$.

There is only one way to make the solution of stability problems truly simple, and that one way is to *ABANDON* Eqs. (11-6) and (11-7). Why? Because the form in which these equations are written seems to rule out the very behavior which usually causes systems to be unstable!

THE BEHAVIOR WHICH USUALLY CAUSES INSTABILITY

One of the keystones of modern engineering is the axiom that "All things flow DOWNhill". Every engineer learns (or should learn) to think in terms of things flowing downhill. A seemingly logical extension of this axiom is "The steeper the hill, the greater the flow rate". Notice that Eqs. (11-6) and (11-7) seem to support this extension--they seem to state that increasing the driving force will increase the flow rate. And therefore in the old engineering, we learn to think that increasing the driving force is a sure way to increase the flow rate.

But it is not always true that increasing the driving force will increase the flow rate. Sometimes increasing the driving force DEcreases the flow rate. Sometimes dynamic processes behave in such a way that, *in a differential sense*, things flow UPhill. This seemingly uphill behavior is usually the cause of instability.

Thermal instability usually indicates that, somewhere in the system, an INcrease in ΔT causes a DEcrease in q.

Fluid flow instability usually indicates that, somewhere in the system, an INcrease in ΔP causes a DEcrease in fluid flow rate.

Because instability generally indicates that something is flowing uphill in a differential sense, Equations (11-6) and (11-7) are of no use in dealing with stability. It is simply not possible to use Eq. (11-6) and envision that an INcrease in ΔT sometimes results in a DEcrease in q. It is simply not possible to use Eq. (11-7) and envision that an INcrease in ΔP sometimes results in a DEcrease in V. This type of behavior is precisely the *opposite* of the behavior Eqs. (11-6) and (11-7) seem to describe.

THE BEHAVIOR SUGGESTED BY THE NEW ENGINEERING CONCEPTS

In the new engineering, we have *NO USE* for Eqs. (11-6) and (11-7). We replace them with

$$q = f\{\Delta T\} \tag{11-8}$$

$$W = f\{\Delta P\} \tag{11-9}$$

Unlike Eqs. (11-6) and (11-7), these fundamental equations of the new engineering are not misleading--they do not seem to describe behavior without actually doing so--they seem to state exactly what they do in fact state.

Eq. (11-8) states that q is an unspecified function of ΔT. This is precisely the statement made by Eq. (11-6), even though it seems to state something quite different.

Eq. (11-9) states that W is an unspecified function of ΔP. This is precisely the statement made by Eq. (11-7), even though it seems to state something quite different.

Eq. (11-8) correctly states that an increase in ΔT may result in either an increase or a decrease in q, depending on $f\{\Delta T\}$. Eq. (11-9) correctly states that an increase in ΔP may result in either an increase or a decrease in W, depending on $f\{\Delta P\}$. And by making it possible to think in this new, free form way, these equations greatly simplify the solution of stability problems in particular and nonlinear problems in general.

THE THERMAL STABILITY OF AN IDEALIZED SYSTEM

The thermal analog of the above fluid flow example is described in the table below:

FLUID FLOW PARAMETER	THERMAL PARAMETER
• Tank level	• Temperature
• Fluid flow rate	• Heat flux
• Fluid flow in/out mismatch results in changing level	• Heat flux in/out mismatch results in changing temperature

We define thermal stability as follows:

> *At a potential steady-state condition, a system is thermally stable if it inherently opposes a small perturbation in temperature at any point in the system and moves toward the unperturbed condition.*

Suppose we wish to analyze the thermal stability of the system shown in Fig. (11-2). This system is analogous to the fluid flow system in the example above because we have made the simplifying assumption that the temperature drop through the wall is small and may be neglected.

Therefore the temperature distribution in the wall is described by a single temperature, just as the tank level was described by a single level. In other words, we are making the simplifying assumption that the thermal behavior of the wall is described by Eq. (11-10):

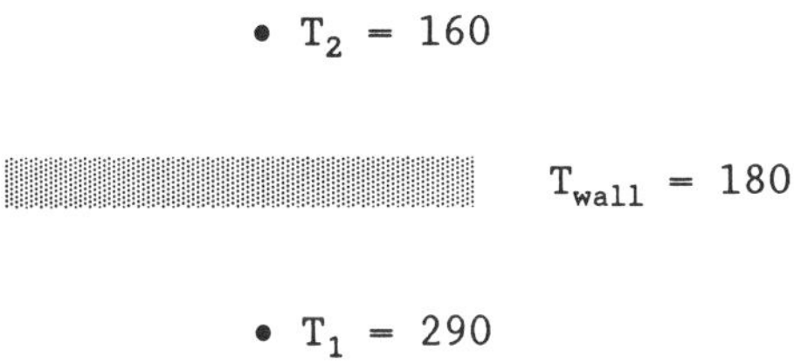

FIGURE (11-2) IDEALIZED THERMAL SYSTEM

$$q_{wall} = -\infty \ dT/dx \qquad (11\text{-}10)$$

The thermal stability analysis of this idealized system requires little more than the recognition that, if q into the wall is greater than q out of the wall, the temperature of the wall will increase, and conversely. Moreover, the behavior of this idealized system closely resembles the behavior of real systems.

As in the fluid flow example, we appraise the stability by introducing a perturbation and analyzing the system's response. We perturb the wall temperature to 180.1 F, and then analyze the system to determine whether or not the perturbation is opposed. The purpose of the analysis is to determine whether or not the mismatch between q into the wall and q out of the wall causes the wall temperature to move toward its initial value. As in the fluid flow example, stability is indicated if the wall temperature moves toward its initial value, and conversely.

As in the fluid flow example, the analysis can be based on either $q\{\Delta T\}$ or $dq/d\Delta T$, but we will generally use the differential form. Now suppose that the functions

$$q_{in}\{\Delta T_{in}\} \qquad q_{out}\{\Delta T_{out}\}$$

are given, and differentiation *at the conditions in Fig. (11-2)* gives:

$$dq_{in}/d\Delta T_{in} = -1 \quad \text{at } \Delta T_{in} = 110 \qquad (11\text{-}11)$$

$$dq_{out}/d\Delta T_{out} = 2 \quad \text{at } \Delta T_{out} = 20 \qquad (11\text{-}12)$$

It is important to notice that these derivatives are evaluated at the conditions given in Fig. (11-2), and therefore they are identical to derivatives with respect to T_{wall} at T_{wall}=180, the wall temperature in Fig. (11-2).

Eqs. (11-11) and (11-12) indicate that, at the conditions in Fig. (11-2),

$$dq_{in}/d\Delta T_{in} < dq_{out}/d\Delta T_{out} \qquad (11\text{-}13)$$

Since the derivatives with respect to ΔT are identical to derivatives with respect to T_{wall} as noted above, we may rewrite Eq. (11-13) in the form

$$dq_{in}/dT_{wall} <s\ dq_{out}/dT_{wall} \qquad (11\text{-}14)$$

If Criterion (11-14) is satisfied, a positive perturbation in wall temperature will result in a decreasing temperature, indicating stability. We therefore conclude that the system in Fig. (11-2) is thermally stable because Eqs. (11-11) and (11-12) satisfy Criterion (11-14). If Criterion (11-14) is not satisfied, we conclude that the system is thermally unstable.

Criterion (11-14) is the thermal stability criterion for this example. Notice that this idealized example is quite generic, and therefore Criterion (11-14) is a simplified form of the generic criterion for thermal stability. The simplification is the fact that the subscript is "wall" rather than an arbitrary location within the wall. (Recall that our definition of thermal stability requires stability with respect to temperature "at any point in the system".)

THE THERMAL STABILITY OF REAL SYSTEMS

Criterion (11-14) does not describe the behavior of real systems. In real systems, the wall in the above example is an important thermal component, and its behavior cannot be approximated by Eq. (11-10). The generic criterion for thermal stability in real systems is:

$$dq_{in}/dT_i <\text{s } dq_{out}/dT_i \qquad (11\text{-}15)$$

where i refers to an arbitrary plane in the system. It is important to note that:

> *Thermal stability requires that Crit. (11-15) be satisfied at ALL planes in the system.*

In general, the "all planes" requirement is met if Criterion (11-15) is satisfied at the plane where the right side of the criterion is a minimum. This plane can usually be determined by inspection as in the example below.

Criterion (11-15) is generic. It applies to *ALL* heat transfer systems. It is the condition which must be met by every thermal system if it is to operate in a thermally stable manner--a steady manner which resists vanishingly small temperature perturbations by inherently generating a force which opposes them.

HOW TO ANALYZE THE THERMAL STABILITY OF A REAL SYSTEM WHICH CONTAINS A "TEMPERATURE CONTROLLED" HEAT SOURCE

The application of Criterion (11-15) is demonstrated in the example below. The system heat source is described by the specification of its temperature, indicating that heat source temperature is independent of heat flux. This type heat source is usually referred to as "temperature controlled". A condensing vapor is a temperature controlled heat source.

In a later chapter, we discuss the analysis of systems which contain "power controlled" heat sources. The analysis of this type of equipment differs because the value of the left side of Criterion (11-15) in large part depends on the characteristics of the heat source power supply. A resistance heated wall is a power controlled heat source.

PROBLEM (11-1):

Appraise the thermal stability of the system shown in Fig. (11-3) for T_1 in the range 100 to 350.

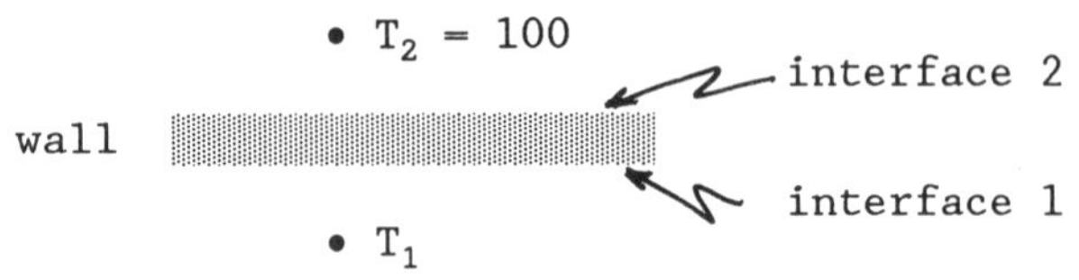

FIGURE (11-3) THERMAL SYSTEM IN PROBLEM (11-1)

INPUT TO PROBLEM (11-1)

$$q_1\{\Delta T_1\} = 6300\ \Delta T_1 \qquad (11\text{-}16)$$

$$q_{wall}\{\Delta T_{wall}\} = 1120\ \Delta T_{wall} \qquad (11\text{-}17)$$

$$q_2\{\Delta T_2\} = 10000\Delta T_2 - 250\Delta T_2^{\,2} + 15.5\Delta T_2^{\,2.5} \qquad (11\text{-}18)$$

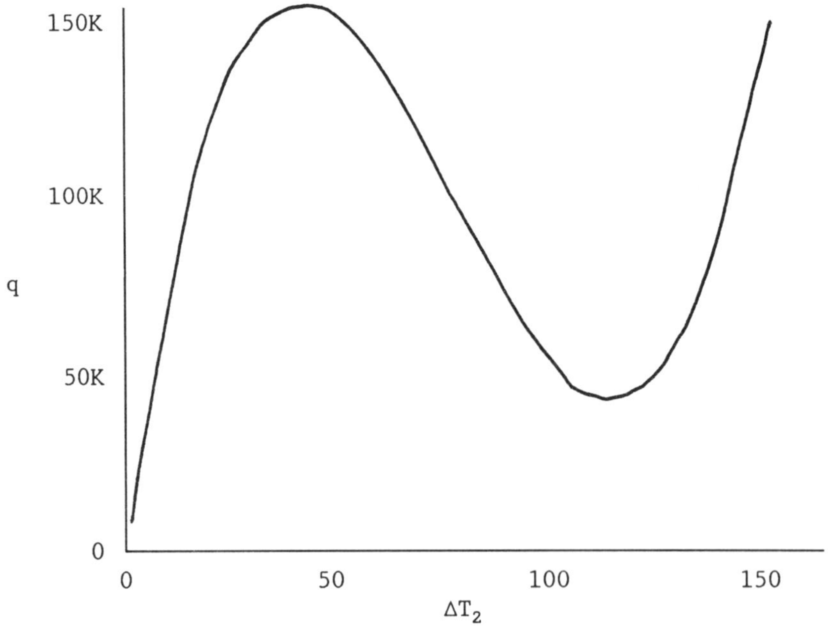

FIGURE (11-4) THERMAL BEHAVIOR OF INTERFACE 2 IN FIGURE (11-3), EQ. (11-18)

ANALYSIS OF PROBLEM (11-1)

The first step in the analysis is to determine whether any wall or interface behaves in such a way that heat is flowing *uphill* in a differential sense--ie in which the $q\{\Delta T\}$ function is such that

$$dq/d\Delta T < 0 \qquad (11\text{-}19)$$

If any wall or interface behaves in the manner described by Expression (11-19), thermal instability is a distinct possibility. If *no* wall or interface behaves in this manner, and if the system heat source is temperature controlled, then thermal stability is assured (because the derivative of q_{in} will be negative, the derivative of q_{out} will be greater than zero, and Crit. (11-15) will therefore be satisfied).

Now let us evaluate the derivatives of the given $q\{\Delta T\}$ functions in order to determine whether any derivative satisfies Expression (11-19):

$$dq_1/d\Delta T_1 = 6300 \qquad (11\text{-}20)$$

$$dq_{wall}/d\Delta T_{wall} = 1120 \qquad (11\text{-}21)$$

$$dq_2/d\Delta T_2 = 10000 - 500\Delta T_2 + 38.75\Delta T_2^{1.5} \qquad (11\text{-}22)$$

The derivatives in Eqs. (11-20) and (11-21) are greater than zero at all values of ΔT. The derivative in Eq. (11-22) is less than zero in the interval

$$39 < \Delta T_2 < 113$$

and greater than zero outside this interval. (The interval is determined by analysis of Eq. (11-22) or inspection of Fig. (11-4).)

Since Criterion (11-15) is obviously satisfied outside the above interval, we conclude that the system is thermally stable outside the interval, and that further analysis is required to appraise thermal stability within the interval.

We noted above that thermal stability requires that Criterion (11-15) be satisfied at all planes in the system. The following applies to systems which contain a temperature controlled heat source and is offered without proof:

> *In systems with only one component which satisfies Exp. (11-19), thermal stability at all planes is demonstrated if Crit. (11-15) is satisfied within the wall or interface which satisfies (11-19).*

Since Exp. (11-19) is satisfied only at Interface 2, we perform the thermal stability analysis there. Using "in" and "out" to refer to the direction of heat flow into and out of Interface 2, and subscript "i" to refer to Interface 2, we write:

$$T_1 - T_i = \Delta T_1 + \Delta T_{wall} \qquad (11\text{-}23)$$

$$T_1 - T_i = q/6300 + q/1120 \qquad (11\text{-}24)$$

$$q = 951(T_1 - T_i) = q_{in}\{T_i\} \qquad (11\text{-}25)$$

$$dq_{in}/dT_i = -951 \qquad (11\text{-}26)$$

$$q_{out}\{T_i\} = q_2\{\Delta T_2\} \qquad (11\text{-}27)$$

$$dq_{out}/dT_i = 10000 - 500(T_i\text{-}100) + 38.75(T_i\text{-}100)^{1.5} \qquad (11\text{-}28)$$

Eq. (11-24) combines (11-23) with (11-16) and (11-17). Eq. (11-28) combines (11-27) and (11-22). Substituting Eqs. (11-26) and (11-28) into Criterion (11-15), we obtain the thermal stability criterion for the subject system:

$$-951 <s\ 10000 - 500(T_i - 100) + 38.75(T_i - 100)^{1.5} \qquad (11\text{-}29)$$

Criterion (11-29) is *not* satisfied in the interval

$$146 < T_i < 203$$

It *is* satisfied outside this interval.

ANSWER TO PROBLEM (11-1)

The system is thermally UNstable at values of T_i in the interval 146 to 203, and is thermally stable outside this interval. Fig. (11-5) summarizes the results of the stability analysis. It describes how both sides of Crit. (11-29) vary with T_i. The shaded region indicates the region of thermal instability.

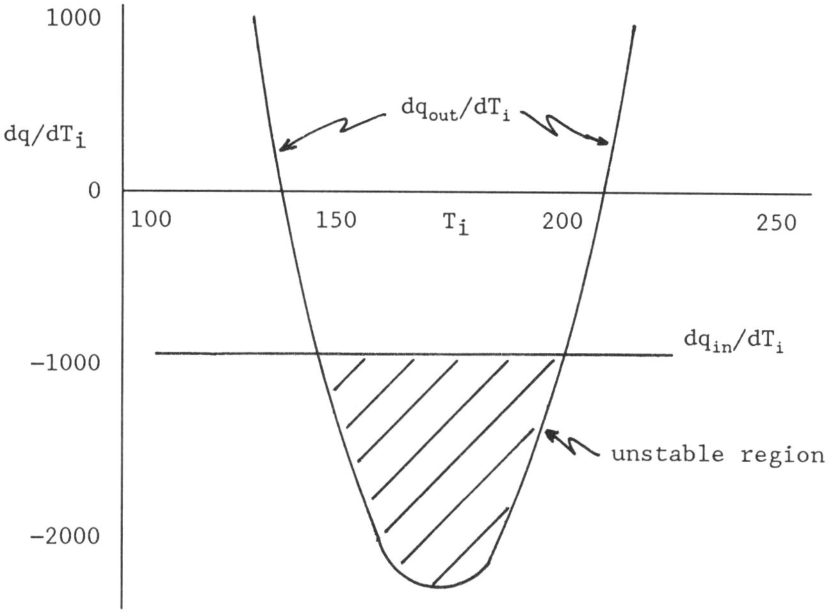

FIGURE (11-5) THERMAL STABILITY ANALYSIS, PROBLEM (11-1)-- dq/dT_i vs T_i

HOW TO DETERMINE THE EFFECT OF THERMAL INSTABILITY ON SYSTEM BEHAVIOR

Problems (11-2) and (11-3) demonstrate how to determine the effect of thermal instability on system behavior. Thermal instability can result in either hysteresis or undamped oscillations. In the problems below, the thermal instability results in pronounced hysteresis. A later chapter demonstrates the analysis of thermal instability which results in undamped oscillations.

PROBLEM (11-2)

Problem (11-1) indicated a region of thermal instability. Describe the effect of the thermal instability on system behavior.

GIVEN

Problem (11-1) and its solution are given.

ANALYSIS OF PROBLEM (11-2)

The best way to describe the effect of thermal instability on system behavior is to determine how the system heat flux or heat load responds to the independent parameter in the heat source, over the complete operating envelope of the equipment. This relationship is called "overall thermal behavior".

If the system heat source is temperature controlled, overall thermal behavior describes how the heat flux responds to the temperature of the heat *source*. If the system heat source is power controlled, overall thermal behavior describes how heat flux responds to whatever heat source parameter is essentially independent--it could be power supply volts or amps, or even the setting of a throttle or a knob.

In Problem (11-1), T_1 is the independent parameter in the heat source, and therefore overall thermal behavior is the function $q\{T_1\}$. This function is determined by the simultaneous solution of Eqs. (11-18) and (11-25):

$$q = 10000(T_i-100) - 250(T_i-100)^2 + 15.5(T_i-100)^{2.5} \qquad (11\text{-}18)$$

$$q = 951(T_1 - T_i) \qquad (11\text{-}25)$$

Because of the nonlinear character of Eq. (11-18), it is simpler to determine $q\{T_1\}$ graphically rather than analytically. This is accomplished in the following way:

- Select a set of T_i values.

- Solve for $q\{T_i\}$ using Eq. (11-18).

- Solve for $T_1\{q, T_i\}$ using Eq. (11-25).

- Solve for dq_{out}/dT_i using Eq. (11-28).

- Appraise Criterion (11-15) at T_i.

- Plot q vs T_1 and note each stability condition. Indicate stable regions by drawing a line between stable operating points. Indicate unstable regions by not drawing a line between unstable operating points. Indicate hysteresis with arrows which indicate the direction of step changes.

The analysis is summarized in Table (11-1). The q vs T_1 plot referred to above is Fig. (11-6).

ANSWER TO PROBLEM (11-2):

The effect of the thermal instability on the behavior of the system in Problem (11-1) is described in Fig. (11-6). Notice in the figure that:

- The effect of thermal instability on system behavior is hysteresis in the T_1 interval 252 to 307. The hysteresis prevents operation at unstable values of T_i--ie at T_i values from 146 to 203.

- As T_1 monotonically increases, q goes through a step decrease at 307 F. The step decrease is approximately a factor of 3. It is accompanied by a step increase in T_i from 146 to 234.

- As T_1 monotonically decreases, q goes through a step increase at 252 F. The step increase is approximately a factor of 3. It is accompanied by a step decrease in T_i from 203 to 121 F.

- Operation is thermally stable at all values of source temperature, even though operation is thermally unstable at certain values of Interface 2 temperature--ie those values which correspond to the unstable points in Fig. (11-6).

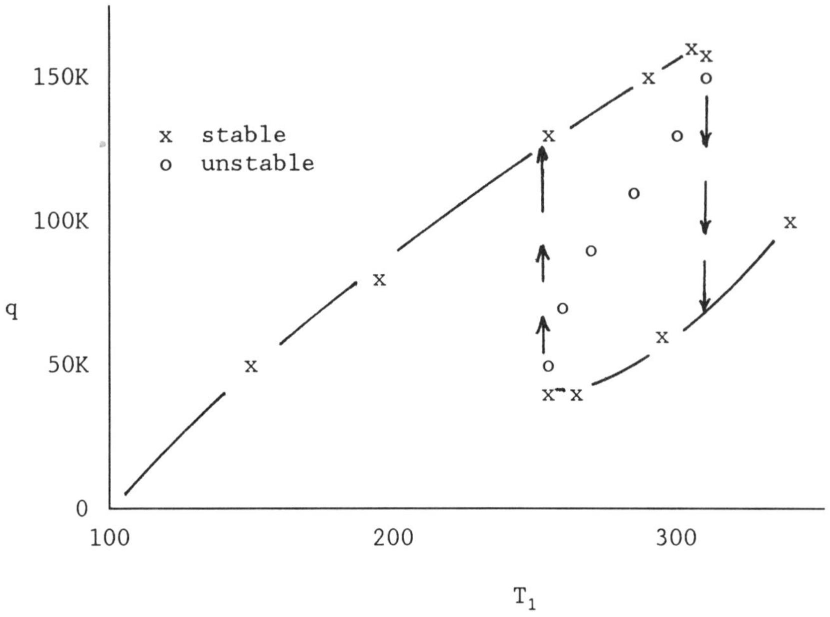

FIGURE (11-6) OVERALL THERMAL BEHAVIOR OF THE SYSTEM IN PROBLEM (11-2) *(Note the hysteresis caused by thermal instability)*

HOW TO ANALYZE THERMAL STABILITY GRAPHICALLY

Thermal stability is a practical problem only in systems which contain a highly nonlinear interface. Because of the cumbersome mathematics often required to analytically describe highly nonlinear behavior, the behavior of such interfaces is usually described graphically. When this is the case, it is much simpler to perform the analysis of thermal stability graphically rather than analytically.

The graphical analysis of thermal stability is illustrated in Problem (11-3).

TABLE 11-1

THE ANALYSIS FOR FIGURE (11-6)

T_i	q from (11-18)	T_1 from (11-25)	dq/dT_i from (11-28)	S or U from (11-15)
102	19100	122	9110	Stable
105	44600	151	7930	S
110	79900	194	6230	S
120	127700	254	3470	S
130	151400	289	1370	S
140	156800	305	-200	S
145	154300	307	-800	S
150	149000	307	-1300	UNSTABLE
160	132200	299	-1990	U
170	110400	286	-2310	U
180	87300	272	-2270	U
190	66100	260	-1920	U
200	50000	253	-1250	U
205	44800	252	-810	STABLE
210	42000	254	-300	S
220	45000	267	940	S
230	61700	295	2440	S
240	94600	340	4190	S

PROBLEM (11-3):

Solve Problems (11-1) and (11-2) graphically using Fig. (11-4) and Eqs. (11-16) and (11-17). Do *not* substitute an analytical expression for the function in Fig. (11-4).

ANALYSIS OF PROBLEM (11-3):

- Select a set of heat source temperatures and determine $q_{in}\{T_1\}$ from Eq. (11-25). For example,

$$q_{in}\{T_1=280\} = 951\{280 - T_i\} \tag{11-30}$$

- Plot Eq. (11-30) and the equations for other values of T_1 on Fig. (11-6). The end result is Fig. (11-7).

- Note the value(s) of q at the intersection(s) of q_{in} and q_{out} in Fig. (11-7).

- Appraise the thermal stability at each intersection--ie solve Crit. (11-15) by inspection of Fig. (11-7). For example, at the middle intersection along the line for T_1=280, the slope of the heat out curve is more negative than the slope of the heat in curve. Therefore Crit. (11-15) indicates that the system is thermally unstable at this intersection, and will not operate in a steady manner at this potential operating condition.

- Determine the effect of thermal instability on system behavior by plotting the $q\{T_1\}$ coordinates for each intersection in Fig. (11-7). The end result is Fig. (11-6) which also was obtained from the analytical solution of Problem (11-1).

ANSWER TO PROBLEM (11-3):

Fig. (11-7) illustrates the graphical solution of thermal stability. It gives the same results as the analytical solution of Problem (11-1). Fig. (11-7) indicates that the system is thermally unstable at T_i values in the interval 146 to 203, and thermally stable at all other values of T_i. Plotting the $q\{T_1\}$ coordinates of the intersections in Fig. (11-7) results in Fig. (11-6).

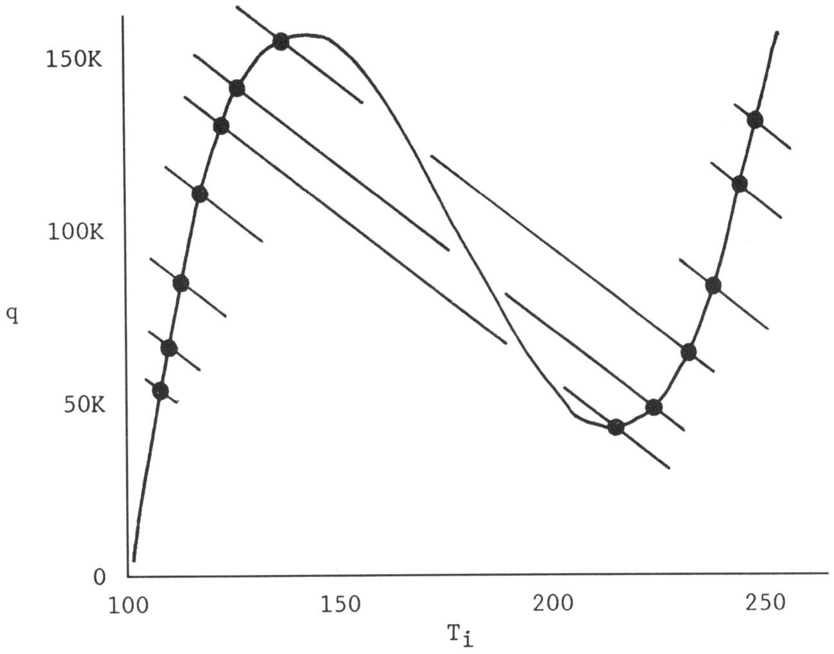

FIGURE (11-7) GRAPHICAL ANALYSIS OF THERMAL STABILITY, PROBLEM (11-3)

THE OLD HEAT TRANSFER VIEW OF THERMAL STABILITY

Problems (11-1) through (11-3) illustrate how to deal with thermal stability using the new heat transfer. The problems deal only with systems which contain a temperature controlled heat source. (A later chapter describes how to analyze the thermal stability of systems which contain a power controlled heat source.)

Today's old heat transfer view of the thermal stability of systems which contain a temperature controlled heat source is described in the following by Rohsenow (1963):

> *With condensing vapor as the heat source on one side of a wall, any point on the entire (pool boiling) curve can be reached under stable conditions.*

This statement reflects the old heat transfer view that thermal stability is assured in systems which contain a temperature controlled heat source such as a condensing vapor. With this heat source type, the old heat transfer supposes that heater surface temperature is an independent variable which can be set at any desired value. Therefore stable operation is assured at any point on the pool boiling curve if the heat source is a condensing vapor.

The above view of thermal stability is generally presented in today's texts on the old heat transfer. For example, Kreith and Bohn (1986) state:

> If the surface temperature is raised further so that the ΔT (across Interface 2) at the maximum heat flux is exceeded, any of three things can occur, depending on the method of heat control and the material of the heating surface (reference to Westwater (1959)):
>
> 1. If the heater surface temperature is the independent variable and the heat flux is controlled by it, the mechanism will change to transition boiling and the heat flux will decrease. . . .

The difficulty with this old view is that the heater surface temperature is *NEVER* the independent variable.

> *The heater surface temperature ALWAYS depends on the thermal coupling (thermal resistance) between the heat source and the surface.*

The cause of the difficulty is that the old heat transfer assigns the independence of the heat source temperature to the heater surface temperature. The old heat transfer view would be correct *ONLY* if the heater surface temperature were the same as the temperature of the heat source--a condition which *NEVER* describes real hardware.

THE SIGNIFICANCE OF PROBLEMS (11-1) TO (11-3)

The system in Problems (11-1) to (11-3) closely resembles the pool boiler referred to in Rohsenow's statement. The heat source in the problem closely resembles a condensing vapor; the thermal behavior of the interface described in Fig. (11-4) closely resembles the widely accepted version of the pool boiling curve.

If the old heat transfer view were correct, the system in Problem (11-1) and all systems like it would operate in a thermally stable manner at all values of interface temperature. However, the analysis of Problems (11-1) to (11-3) *rigorously* demonstrates that the system is thermally UNstable over a considerable interval of interface temperature, and therefore the old heat transfer view of thermal stability is simply incorrect.

If the old heat transfer view of thermal stability were correct, the following would be true:

- The system in Problem (11-1) would be thermally stable at all values of T_i.
- All the answers arrived at in Problems (11-1) to (11-3) would be incorrect. There would be no hysteresis--no threefold step changes in the value of q--no 100 F step changes in the value of T_i.
- The analysis of Problems (11-1) to (11-3) would not be necessary. The analysis would involve nothing more than inspection of the heat source to determine whether it is temperature controlled or power controlled.

The new heat transfer analyses and solutions of Problems (11-1) to (11-3) are simple and obviously correct. These problems provide the definitive proof that the new heat transfer effectively deals with the thermal stability of temperature controlled pool boilers, and today's old heat transfer does not.

CHAPTER 12

CONCEPTS

Concepts are intellectual tools which make it possible to understand/summarize/order/predict the behavior of Natural phenomena. Concepts are created by intellectually generating a view of Natural phenomena which seems to accord with experiment--which seems to agree with the available data.

The four elements--earth, air, fire and water--seemed to agree with the data on matter. "Spontaneous generation" seemed to agree with the available data on the origin of life. "Phlogiston" seemed to agree with the data on fire. "Caloric" and the "electric fluids" seemed to agree with the data on heat and electricity.

These old concepts were not passing fancies. They were not held tentatively with one eye on the horizon, expectantly watching for better concepts. They were widely regarded as scientific facts which would last forever, and many did in fact survive for *centuries*.

Yet in spite of their long and universal acceptance, all these old concepts have been replaced by concepts which better agree with the available data. And if history is a reliable guide, then the better concepts of today will *surely* be replaced by even better concepts tomorrow.

Today's engineering science is founded on concepts which have been universally accepted for almost 200 years. Ohm's Law and electrical resistance. "Newton's Law of Cooling" and h. Young's Law and moduli. These mature concepts are the keystones of today's engineering science. They are not passing fancies. They are scientific *facts* not open to question--scientific facts which will live *forever*.

But we *must* question today's concepts because *all* concepts are products of the intellect, and the intellect has *always* been unreliable. We must question them because if we do not question them, we have *no hope* of finding better concepts. And we should be confident that better concepts are possible because the history of science demonstrates that there have *always* been better concepts.

Before we question today's engineering concepts, let us use this chapter to consider the nature and purpose of scientific concepts. Our goal is to gain increased objectivity by attaining a better appreciation of what concepts are and what they are not--of how they are generated and what they are intended to accomplish.

Then in the following chapters, we will seriously question today's engineering concepts. And we will consider the concepts of the "new engineering", an engineering science which bears little resemblance to today's engineering because it rejects/replaces many of its principal concepts.

THE TRANSFORMATION OF CONCEPTS

Time has a way of transforming concepts. When a concept is in its infancy, it is appropriately viewed as a state of developing art--an intellectual and loosely held view of Nature which is accepted for only one reason--because it seems to agree with the available data.

But when a concept has been generally accepted for generations, it loses its intellectual and fallible character. It ceases to be a *view* of Nature, and it becomes a *part* of Nature--a scientific *fact*.

For example, after caloric had been in general use for some time, it ceased to be a way to view heat flow phenomena, and became a scientific fact. Caloric began to actually exist--the experiments proved it! Caloric was not an intellectual invention. Caloric was real! And if this seems incredible, we have Maxwell's (1883) testimony that caloric was in fact regarded as a real substance:

. . . there was no doubt in the minds of many eminent men that caloric was a fluid pervading all bodies. . .

And of course the point is that we must be careful not to transform concepts from what they are to what they are not:

- Concepts are not reliable/permanent *facts* of Nature. Concepts are fallible/temporary *views* of Nature.
- Concepts are not discovered--they are *invented.*
- Concepts are not found in Nature--they are found only in the intellect.
- Concepts are not *from* Nature--they are *about* Nature.

Because concepts are unreliable products of the intellect, we should always question the current concepts and be ready/willing/anxious to abandon them whenever better concepts are conceived.

FUNDAMENTAL AND CONTRIVED CONCEPTS AND PARAMETERS

Today's engineering concepts and parameters may be classified into two groups:

- Fundamental concepts and parameters.
- Contrived concepts and parameters.

Fundamental concepts are independent of other concepts. For example, temperature is a fundamental concept. It in no way depends on other concepts for its existence. It is a basic/primordial/free-standing idea used to quantify Natural phenomena.

Contrived concepts are derived from other concepts, and therefore they can be defined only in terms of other concepts. Contrived concepts may be a desirable addition to fundamental concepts, but they are obviously unnecessary because they are readily replaced by the concepts from which they are contrived.

The heat transfer coefficient is an example of a contrived concept. It is contrived from the primary concepts heat and temperature. A rigorous definition of h requires the fundamental concepts heat and temperature. h is in fact a shorthand alias for "the ratio of q to ΔT". This shorthand notation may be convenient, but it is obviously not necessary. If we are truly interested in this ratio, we can simply write $q/\Delta T$ instead of h.

Fundamental parameters are found in Nature. Contrived parameters are *not* found in Nature--they are found only in the intellect. Temperature and heat are examples of fundamental concepts/parameters. They are found in Nature and can be measured. h is a secondary concept/parameter. h is not found in Nature and can not be measured. h can be determined only by measuring q and ΔT and then calculating the ratio $q/\Delta T$.

The main difference between old and new engineering lies in the role played by contrived concepts:

> *Contrived concepts such as h are of primary importance in today's old engineering.*
>
> *Contrived concepts such as h are of NO IMPORTANCE in the new engineering. They are ABANDONED!! They are replaced by the fundamental parameters from which they are contrived.*

The new engineering rejects the contrived concepts of the old engineering because they are unnecessary and undesirable. They are unnecessary because they can be replaced by the fundamental parameters from which they are contrived. They are undesirable because many of them combine the variables we most want to separate, thereby complicating and confusing the solution of nonlinear problems.

The transition from old to new engineering requires that contrived parameters be *abandoned*. They are replaced by fundamental parameters used in a new way--a way which keeps them *separate*.

Because the fundamental parameters are separate in the new engineering whereas they are combined in the contrived concepts of the old engineering, it becomes possible to solve engineering problems with separated variables. And for that reason, the solution of nonlinear problems is much simpler in the new engineering.

CONCEPTS AND DATA

> *In both the old and the new engineering, the purpose of experiment is to measure the value of the parameters involved in the experiment--to obtain the data.*

In today's old engineering, the word "data" refers to the values of fundamental parameters measured in the experiment, *AND* to the values of contrived parameters obtained by manipulating measured values.

In the new engineering, the word "data" refers *ONLY* to measured values. We insist that values obtained by manipulating measured values are results, not data. We use the word "nondata" to refer to values obtained by manipulating measured values of fundamental parameters.

For example, suppose we perform a simple heat transfer experiment in the old engineering and we tabulate the data as in Table (12-1).

Are all the values in Table (12-1) data?

$T_{surface}$	T_{fluid}	q	h
211	199	36	3.0
218	201	58	3.4
229	203	99	3.8
238	204	139	4.1

TABLE (12-1) DATA SET, OLD HEAT TRANSFER

In the old heat transfer, the answer is "Yes." But in the new heat transfer, the answer is an emphatic "No!" The h values are not data. The h values are *nondata*--they are the result of manipulating the measured values of q and ΔT--of dividing q by ΔT.

Heat transfer coefficients cannot be measured in experiments because heat transfer coefficients do not exist in Nature. Heat transfer coefficients "exist" only in the intellect, and only because convective heat transfer is viewed through the cloudy lens inappropriately called "Newton's Law of Cooling".

NONDATA

In the new heat transfer, we distinguish between fundamental parameters and contrived parameters, and we distinguish between data and nondata. Data refers to the measured values of fundamental parameters. Nondata refers to the values of contrived parameters obtained from measured values of fundamental parameters. We consider that:

- Nondata is not Nature's infallible testimony. Nondata is intellectual invention.

- Since nondata is not Nature's testimony, there is no good reason to suppose that nondata has lasting value.

- Since nondata may not have lasting value, we should always insist on reporting/archiving data rather than nondata.

And we should be ready/willing/anxious to reject the concepts which underlie nondata as soon as better concepts are conceived.

A PREVIEW OF THE NEXT FEW CHAPTERS

In the next few chapters, we question the usefulness of today's engineering concepts. We discuss the concepts of the "new engineering", an engineering science which bears little resemblance to today's engineering science because it has nothing to do with contrived concepts. They are replaced by new and different concepts which result from a new and different way of generating concepts.

The next few chapters describe the rationale which underlies the concepts of the new engineering:

- The new engineering is based on a new method of generating concepts. The application of this new method to engineering science results in the rejection and replacement of many concepts and equations which are fundamental in today's engineering For example, the new engineering rejects "Ohm's Law".

- Ohm's Law is replaced with a free form expression which permits the functional relationship between volts and amps to assume all forms including strictly proportional and highly nonlinear. (Just as the free form expression which replaces "Newton's Law of Cooling" permits all forms of functional relationships between q and ΔT.)

- The new engineering rejects the idea of "electrical resistance" in particular and the idea of "resistance" in general. The idea of "resistance" is replaced by the idea of "behavior". This new idea is even more pervasive in the new engineering than resistance is in today's old engineering.

- The new engineering rejects *ALL* of the resistance parameters of today's engineering sciences. It has nothing to do with electrical resistance, diffusive resistance, hydraulic resistance, pneumatic resistance, etc. Resistance parameters are not replaced because they are not necessary, nor have they ever been necessary.

The new engineering sciences are better because they are simpler than today's engineering sciences. *Much* simpler.

CHAPTER 13

OHM'S LAW AND TODAY'S OLD ELECTRICAL SCIENCE

Electrical science was a true science in the 19th century. Ohm's Law described the behavior of conductors in a general way, and "electrical resistance" uniquely described a concept and a parameter.

At the turn of the century, the advent of nonlinear devices meant that the electrical science based on Ohm's Law was no longer a true science because Ohm's Law no longer described the behavior of conductors in a general way. The invention of nonlinear devices made it necessary either to abandon Ohm's Law, or to revise it so that it would again describe the behavior of conductors in a general way.

Today's old electrical science is *NOT* the result of abandoning or revising Ohm's Law. It is the result of adding a second electrical science and several types of resistance to 19th century electrical science. It is described by the following:

- Today's electrical science is based in part on Ohm's Law even though it is *NOT* a Law--even though it does *NOT* describe the behavior of conductors in a general way.

- Today's electrical science is not a science. It is *two* sciences--Ohm's Law science, and non-Ohm's-Law science.

- In today's electrical science, resistance is not a concept--it is *several* concepts.

The proper goal of science is logical simplicity. It is neither logical nor simple to have a Law which is not a Law--a science which is two distinct sciences--a concept which is several unrelated concepts. Today's electrical science is neither logical nor simple. It is open to serious question, and we question it.

THE CONCEPT OF ELECTRICAL RESISTANCE

Electrical resistance (R) is one of the fundamental concepts of twentieth century engineering science. It was born almost 200 years ago, and it is today widely regarded as a permanent, scientific fact not open to question.

It would be quite accurate to bring Maxwell's observation on caloric into the twentieth century with the following paraphrase:

> *. . . there is no doubt in the minds of many eminent men that electrical resistance is real.*

Electrical resistance is not an invention of the intellect. It is *real*! To doubt the existence of electrical resistance is to doubt experience and common sense. Electrical resistance is here to stay--*forever!*

> *But if caloric seemed real, isn't it possible that electrical resistance only SEEMS real?*

If electrical resistance is real, it will last forever. But if it is not real--if it is merely an intellectual invention in the manner of caloric--then there is no good reason to suppose that it will last forever. In fact, if electrical resistance is an intellectual invention, there is good reason to suppose that it will *NOT* last forever.

How can we determine whether electrical resistance actually exists? It certainly seems to exist. If it does not exist, how can it have lasted for almost 200 years? If it does not exist, how can it be so useful? How can it be used all over the world, and with such fantastic results, if it does not exist--if it is not Nature's way?

We are poorly qualified to make a judgement about the existence or nonexistence of electrical resistance. We are eight generations removed from the birth of electrical resistance, and we are prejudiced by the knowledge that it has been successfully and universally used for a very long time. If we could simply go back eight generations in order to become Ohm's contemporaries, we would be much better qualified to make a judgement about the existence of electrical resistance.

MAXWELL'S VIEW OF ELECTRICAL RESISTANCE

Since we are poorly qualified to make a judgement about the existence of electrical resistance, let us consider the judgement made a century ago by Maxwell (1873)--a giant of electrical science. The following quote is Maxwell's reply to the question "Exactly what is electrical resistance?":

> *OHM'S LAW*
>
> *The electromotive force acting between the extremities of any part of a circuit is the product of the strength of the current, and the resistance of that part of the circuit.*
>
> Here a new term is introduced, the resistance of a conductor, which is defined to be the ratio of the electromotive force to the strength of the current which it produces. The introduction of this term would have been of no scientific value unless Ohm had shown, as he did experimentally, that . . . it has a definite value which is altered only when the nature of the conductor is altered.
>
> In the first place, then, the resistance of the conductor is independent of the strength of the current flowing through it. . . .
>
> The resistance of a conductor may be measured to within one ten thousandth, . . . and so many conductors have been tested that our assurance of the truth of Ohm's Law is now very high.

(Ohm (1827) did not use the term "resistance" in his "Law of the Current" which is now known as Ohm's Law. Ohm used the phrase "reduced length" (a reference to the equivalent length of wires) in place of "resistance" which came into use a short time later.)

(Notice that the heat transfer analog of Maxwell's statement is: "h is the ratio of q to ΔT. This ratio has no scientific value unless it is independent of q." How do you suppose Maxwell would react if we told him that, in the twentieth century, h is used even though its value oftentimes depends on q--even though it oftentimes has "no scientific value"?)

In Maxwell's view, electrical resistance is not a parameter found in Nature--it is not a scientific fact. It is merely a "term" introduced by Ohm--a term "defined to be the ratio of electromotive force to the strength of the current which it produces". In other words, "resistance" does not truly exist. "Resistance" is nothing but a shorthand way to write "the ratio of volts to amps"! And "R" is nothing but a shorthand way to write V/I!

Notice that R is a contrived parameter, as evidenced by the fact that Maxwell does not define it in an absolute way, but rather defines it in terms of electromotive force and current.

The idea of electrical resistance has scientific value to Maxwell because it is a *constant* ratio, and because "so many conductors have been tested that our assurance of the truth of Ohm's Law is now very high". In other words, Ohm's Law is a true Law because it describes electrical behavior in a *general* way.

Wheatstone (1843) also expressed the view that Ohm's Law is valuable because it describes behavior in a general way:

> Viewing the law of the electric circuit from the point at which the labors of Ohm has placed us, there is scarcely any branch of experimental science in which so many and such various phenomena are expressed by formulae of such simplicity and generality.

In summary, Maxwell and Wheatstone accepted Ohm's Law and the concept of electrical resistance because they described electrical behavior in a general way. But don't they implicitly state that they would *reject* Ohm's Law and the concept of electrical resistance if/when they ceased to describe electrical behavior in a general way?

MAXWELL'S QUESTION

Imagine that we bring Maxwell forward to 1989, and that we give him the following brief description of today's electrical science:

> Mr. Maxwell, we now have many conductors which do *NOT* obey Ohm's Law. Many conductors behave in such a way that the "ratio of electromotive force to the strength of the current it produces" is *NOT* "independent of the strength of the current flowing through it". As you yourself noted, Ohm's concept of resistance has "*NO* scientific value" for these conductors.
>
> These "non-Ohm's-Law" conductors have immense practical importance today. Much of today's technology depends on the fact that many conductors do *NOT* obey Ohm's Law.

How do you suppose Maxwell would react to this information? Doesn't it seem likely that his first reaction would be utter amazement? That his second reaction would be to assume we had rejected Ohm's Law and R--just as he indicated he would if/when they ceased to describe electrical behavior in a general way? And doesn't it seem likely that his third reaction would be to ask:

> *What do you use in place of Ohm's Law and resistance?*

THE TWENTIETH CENTURY ANSWER TO MAXWELL'S QUESTION

Within the framework of today's electrical science, the answer to Maxwell's question is:

Ohm's Law and electrical resistance have not been replaced. They are still an important part of today's electrical science. But we now recognize that Ohm's Law is not a universal Law. Ohm's Law applies only if it applies. And quite often, it does not apply.

There are now *two* electrical sciences:

- Ohm's Law science, used when Ohm's Law applies.
- Non-Ohms-Law science, used when Ohms Law does not apply.

The concept of electrical resistance has been greatly expanded. There are now two large classes of conductors:

- Conductors which "obey" Ohm's Law. They behave in a proportional way and are called Ohm's Law resistors or ohmic resistors or linear resistors.

- Conductors which do not obey Ohm's Law. They behave in a nonlinear way and are called non-Ohm's-Law resistors or nonohmic resistors or nonlinear resistors or semiconductors.

There are now several types of electrical resistance:

- *Ohm's Law resistance*. This is the resistance with which you are familiar. It is defined to be the constant ratio of V to I, or equally the proportionality constant between V and I. It is often called static resistance.

- *Dynamic resistance*. This is the derivative of V with respect to I. It is generally used when the value of R depends on I--ie when the value of V/I depends on I. It is a point value used over a narrow range in the vicinity of an operating point. It is a linear approximation of nonlinear V{I} behavior, and is often called incremental resistance or small-signal resistance.

  Static and dynamic resistance are the most important types of resistance. Notice that static resistance is V/I, whereas dynamic resistance is $\Delta V/\Delta I$ as $\Delta \rightarrow 0$.

- *Forward resistance*. This also is a linear approximation of nonlinear V{I} behavior. It differs from dynamic resistance in that the approximation is made over a wide range, and it is used over a wide range.

- *Negative resistance*. This type of resistance should really amaze you. It results when the V{I} behavior is so highly nonlinear that it contains a region in which the derivative of V with respect to I is *negative*--ie a region in which the current INcreases when the voltage drop DEcreases! In a differential sense, the current acts as though it is flowing UPhill! Imagine using Ohm's Law to cope with negative resistance!

Negative resistance is closely analogous to the heat transfer phenomenon called "transition boiling". In this type of boiling, the heat flow DEcreases as the temperature difference INcreases.

Volts and amps are important in both Ohm's Law science and non-Ohm's-Law science, but of course the ratio of volts to amps is not constant in non-Ohm's-Law science. We usually distinguish between the two types of resistors by the terms "linear" and "nonlinear". (But it would be more fitting to use the terms "proportional" and "nonproportional", since Ohm's Law resistance deals effectively only with behavior which is strictly proportional.)

The "resistance" parameter is used in both sciences, but we no longer agree on exactly what resistance is.

- Some of us define "resistance" to be the constant ratio V/I, or the proportionality constant between V and I. Thus resistance is defined only if the ratio V/I is a constant. Since resistance is a constant, the equation

  $$V = IR \tag{13-1}$$

  indicates that V is proportional to I. In this case, Eq. (13-1) is referred to as "Ohm's Law". (White (1962) offers this view of Ohm's Law and resistance.)

- Some of us define resistance to be the variable ratio V/I. Thus resistance is defined even if the ratio V/I is variable. Since R is variable, the equation

  $$V = IR \tag{13-2}$$

  is *NOT* Ohm's Law and R must be treated as a variable in the equation. Eq. (13-2) is not a phenomenological description of electrical behavior. It is merely the defining equation for R. In this case, "Ohm's Law" is the observation that resistance is independent of current for a certain class of resistors. (Halliday and Resnick (1978) offer this view of Ohm's Law and resistance.)

(Notice that the first interpretation is analogous to Fourier's view that h is a constant of proportionality. The second interpretation is analogous to today's view that h is the variable ratio $q/\Delta T$.)

In other words, some of us view electrical resistance as a constant ratio which is defined only for components which exhibit proportional behavior. Others view electrical resistance as a variable ratio which is defined for all components and all types of behavior. There are serious difficulties with both views:

- If electrical resistance is a constant ratio, it is inconsistent to have several types of resistance which are NOT constants and are used in non-Ohm's-Law science.

- If electrical resistance is a variable ratio, then "it has no scientific value" as Maxwell noted.

It is difficult to discuss electrical resistance because the phrase "electrical resistance" refers to several different ideas/parameters. In order to avoid ambiguity, let us agree on the following:

- The symbol R refers to the constant ratio V/I. It denotes that V/I is independent of I--ie that V and I are proportional.

- R is defined only if the ratio V/I is constant--only if V/I is independent of I. Therefore Ohm's Law is given by V = IR.

- V/I refers to the ratio of V to I in a general way. It denotes that V/I may depend on I--ie that V/I may be a variable and therefore it must be treated as a variable.

We have little use for V/I when dealing with non-Ohm's-Law resistors--ie when the ratio of V to I depends on the current. As you know and often stated, if the value of V/I (which you called "resistance") depends on the current, resistance has "no scientific value".

I think you will agree there is no exaggeration in saying there are now two electrical sciences. In fact, electrical science is usually taught as two distinct sciences. Ohm's Law science is taught at an elementary level in all technical programs. Non-Ohm's Law science is taught at a higher level, and to a much smaller group.

In summary, the answer to your question is:

- Ohm's Law science has not been replaced. It has been supplemented with a second electrical science. In this second science, volts and amps are *OFTEN* related in a nonlinear way which little resembles Ohm's Law.
- Ohm's Law resistance has not been replaced. It has been supplemented with several new types of electrical resistance which are NOT equal to "the ratio of electromotive force to the strength of the current it produces".

MAXWELL'S RESPONSE TO TODAY'S OLD ELECTRICAL SCIENCE

My colleagues and I accepted Ohm's Law and its concept of electrical resistance because, in our time, they described behavior in a general way. I was certain that Ohm's Law and resistance would be abandoned if scientific progress should ever rob them of their generality. Yet you do not reject them even though they lack generality in your time!

I am somewhat discomfited by the fact that you still refer to Ohm's "Law" in spite of the fact that it has not been a Law for almost a century. Wouldn't it be more fitting to refer to it as "Ohm's Rule of Thumb" in order to indicate that it lacks generality?

If electrical "resistance" is no longer a singular concept, does it still have scientific value? If electrical "resistance" does not uniquely identify a parameter, is the idea of resistance still useful? Perhaps it would be better to abandon resistance rather than multiply it. Have you considered abandoning the concept/idea/parameter you call electrical resistance? Or rather, have you considered abandoning the various concepts/ideas/parameters you call electrical resistance?

To be quite frank, I intensely dislike the idea of two distinct electrical sciences--one to deal with proportional behavior, the other to deal with nonproportional behavior. The proper goal of science is logical simplicity, and it is neither logical nor simple to have two electrical sciences.

I do not mean to be inappreciative of your science, but it seems to me there must be a better way--a simpler way--to construct an electrical science which deals effectively with twentieth century electrical phenomena. And surely that better way involves abandoning Ohm's Law and resistance, since they no longer describe electrical behavior in a general way.

THE NEW ENGINEERING RESPONSE TO MAXWELL'S RESPONSE

There is in fact a new electrical science which abandons Ohm's Law and the various concepts/ideas/parameters of electrical resistance. The new electrical science is part of the "new engineering", a new science which abandons *ALL* the Laws and resistance concepts of today's engineering, and is based on the new and simple idea that:

> *Engineering sciences should be founded on free form descriptions of behavior, using fundamental parameters such as temperature, pressure, flow rate, volts, amps, etc. They should have NOTHING TO DO with contrived parameters such as electrical resistance, heat transfer coefficient, friction factor, etc.*

For example, we should deal with electrical phenomena using $V\{I\}$ instead of the ratio V/I called electrical resistance. We should deal with convective heat flow phenomena using $q\{\Delta T\}$ instead of $q/\Delta T$ called heat transfer coefficient. We should deal with incompressible fluid flow phenomena using $W\{\Delta P\}$ instead of a resistance parameter which is proportional to $\Delta P/W^2$ and is called friction factor.

The next several chapters summarize the new engineering and illustrate its application to practical problems. The treatment is not comprehensive, but it should give you a good idea of what it is like and why it is better. *Much* better.

CHAPTER 14

THE NEW ELECTRICAL SCIENCE

THE NEW ENGINEERING ANSWER TO MAXWELL'S QUESTION

Mr. Maxwell, Ohm's Law and its "resistance" concept have been replaced because they no longer describe electrical behavior in a general way. Ohm's Law has been replaced by a new Law which has nothing to do with resistance, and is obeyed by all electrical devices. In terms which resemble your 1873 definition of Ohm's Law, the new Law is:

> The electromotive force acting between the extremities of any part of a circuit is a function of time and the current in that part of the circuit. The functionality between emf (V, volts), current (I, amps), and time (t, secs) is not prescribed, and all functions are allowed. The V{I,t} functionality is referred to as "electrical behavior". Symbolically, the new Law is:
>
> $$V = f\{I,t\} \qquad (14\text{-}1)$$
>
> The electrical behavior of many devices is independent of time, and their behavior is more simply described by
>
> $$V = f\{I\} \qquad (14\text{-}2)$$
>
> They are called V{I} devices. (In today's electrical science, they are called "ideal resistors", indicating that they do not exhibit "inductance" or "capacitance", and that emf and current are independent of time.)

At first glance, this new way of dealing with electrical phenomena may seem a step backward. The new "Law" hardly seems a Law at all, since it allows all forms of functionality. But if you consider the rationale behind the new

Law, you will come to recognize that it results in a new electrical science which is much better than the old.

THE RATIONALE BEHIND THE NEW LAW

In the twentieth century, we have many electrical devices which do not exhibit the proportional electrical behavior with which you are familiar. The relationship between emf and current is now infinitely varied, and includes all forms of behavior from proportional to highly nonlinear.

Because of this infinite variety, we require a Law which allows every type of functional relationship. The new Law must state what is related to what, but it must not dictate functionality because no single form describes electrical behavior in a general way. In short, the new Law must be free form.

Notice that both Eqs. (14-1) and (14-2) satisfy these requirements. They state what is related to what, and they do not dictate or even suggest the form of the relationship. Eq. (14-1) is the general form, and it applies to all devices. Eq. (14-2) is a particular form of Eq. (14-1), and it applies only to a specific type of device.

Eq. (14-1) is *exactly* the type of equation we want and need to deal with twentieth century devices--*ALL* twentieth century devices. And since this free form equation allows all forms of behavior, it will be adequate to deal with all future devices, no matter how they behave.

> *Because Eq. (14-1) is exactly the expression we want and need to deal with ALL current and future devices, it is the fundamental Law for the new electrical science.*

THE REPLACEMENT FOR THE PARAMETER "R"

In the new electrical science, there is no replacement for the resistance parameter R (ohms) because R is unnecessary and has always been unnecessary.

If this seems implausible, recall that your own definition of R emphasizes that R has no fundamental character--R is simply a "term" which you substitute for "the ratio of emf to current". The old parameter "R" is nothing more than a shorthand way to write V/I. Therefore it is self evident that the parameter/concept/idea "R" is unnecessary and has always been unnecessary.

Electrical resistance is not *THE* way to deal with resistive electrical phenomena. It is merely *ONE* way. It is a way of dealing with V and I using the ratio V/I. This ratio is viewed as a resistance to the flow of electricity, and is christened "electrical resistance".

> *But no matter how we view it or what we call it, resistance is simply the ratio of V to I. And R is simply a shorthand way to write V/I.*

This ratio or resistance method works well only if V/I is independent of I--only if the "resistance" is independent of the current--only if V and I are proportional. As you so aptly put it, this method is "of no scientific value" if the value of V/I depends on I.

Electrical behavior is another way of dealing with resistive electrical phenomena. It is a way of dealing with V and I using V{I} instead of V/I--using behavior instead of resistance. The function V{I} is christened "electrical behavior" because V{I} *is* electrical behavior.

When dealing with resistive electrical phenomena, Eq. (14-2) and its behavior concept work well *no matter how V and I are related*. In your terminology, Eq. (14-1) and electrical behavior have "scientific value" even if the "resistance" depends on the current.

Ohm's Law is a particular form of Eq. (14-2). If Ohm had conceived Eq. (14-2) and deduced its application, it would have been the foundation of electrical science in the 19th century. And the concept of electrical resistance would not have been invented.

In the new electrical science, we do not replace R because R has always been unnecessary, and because we have no interest in the ratio of emf to current. If some day we should become interested in this ratio (a prospect which seems most unlikely), we will simply write V/I.

THE REPLACEMENT FOR THE IDEA OF ELECTRICAL RESISTANCE

The concept of "resistance" is an idea as well as a parameter. The generic idea is that flow requires a driving force in order to overcome the Natural forces which are imagined to oppose all flows. "Resistance" is the result of these imagined forces.

In the new engineering, we recognize that "resistance" is a way to think about flow phenomena, just as caloric is a way to think about heat flow. Resistance is an invention, just as caloric is an invention. Resistance is not Nature's way. It is *man's* way. Man's way to view Nature--man's way to interpret Nature's way--man's way to predict the behavior of Natural phenomena.

Man *invented* the idea of electrical resistance--the idea that the ratio V/I can be viewed as a "resistance" which opposes the flow of electricity. As you once noted, this idea of electrical "resistance" is of "no scientific value" unless "the resistance of the conductor is independent of the strength of the current flowing through it". In other words, the idea that V/I can be viewed as resistance is of no scientific value unless V/I is independent of I--unless V is proportional to I.

You accepted the idea of electrical resistance because V/I was generally independent of I in your time. But in our time, V/I is *NOT* generally independent of I, and therefore the idea of resistance now has *NO scientific value*.

Because the idea of electrical resistance no longer describes electrical behavior in a general way, it no longer has scientific value. For this reason, it is rejected in the new electrical science.

DEALING WITH NONLINEAR BEHAVIOR

In today's electrical science, we deal with nonlinear devices by inventing an additional science and additional resistances. But in the new electrical science, we deal with nonlinear devices by rejecting *all* forms of resistance/conductance. And we use *ONE* science to deal with *ALL* electrical devices.

In today's Ohm's Law science, we focus our attention on V/I which we call "electrical resistance". In the new science, we have no interest in electrical resistance--no interest in V/I. We focus our attention on V{I} which we call "electrical behavior".

We abandon the words resistance, conductance, capacitance, and inductance. We replace the words resistor, conductor, capacitor, and inductor with the word device or component. We retain the words resistive, conductive, inductive, and capacitive with their current meanings. We often use V{I} to denote resistive/conductive, and V{I,t} to denote inductive/capacitive.

In the new engineering, we view flow phenomena using the generic idea that flow responds to driving force, and we concentrate on the function "flow {driving force}". This idea is found throughout the new engineering, just as the idea of resistance is found throughout the old engineering. For example:

- If the analysis concerns resistive electrical flow devices, we concentrate on electrical behavior--on the behavior of volts vs amps--on V{I}.

- If the analysis concerns heat flow phenomena, we concentrate on thermal behavior--on the behavior of heat flux vs thermal driving force--on q{TDF}.

• If the analysis concerns liquid flow phenomena, we concentrate on liquid flow behavior--on the behavior of liquid flow vs pressure difference--on W{ΔP}.

THE PARTICULAR ADVANTAGE OF THE NEW ELECTRICAL SCIENCE

The new electrical science is *ONE* science, and it deals in a simple and effective way with *ALL* types of electrical behavior. It is a logical and simple science, and is based on only two electrical parameters--V and I.

Ohm's Law science and Ohm's Law resistance can deal effectively only with proportional behavior, and therefore it was necessary to invent an additional electrical science and additional resistance parameters in order to deal with the nonlinear resistors of the 20th century. It is neither logical nor simple to have two electrical sciences and several unrelated resistance parameters.

MAXWELL'S RESPONSE TO THE NEW ELECTRICAL SCIENCE

"What you say sounds important and elegant, but I fail to see how it can be used in a practical way.

Your new Law states that volts and amps are related.
What's new about that?

You find fault with Ohm's Law because it cannot be used for all types of resistors/conductors, but at least it can be used for one type! The new Law makes good rhetoric, but I fail to see how it can be used for even one resistor.

I am inclined to agree with you that two electrical sciences is not aesthetically pleasing, but obviously it has been very successful. Why change it?

In short, I don't like the new Law because I don't see how it is to be used--or even that it *can* be used!

If you want to convince me that the new electrical science is better than the two current electrical sciences, you will have to show me how to use the new science to solve practical problems.

THE RESPONSE TO MAXWELL'S REQUEST

Mr. Maxwell, in response to your request, the next chapter describes how to use the new electrical science to solve problems which deal with purely resistive circuits. By way of preview, the problems illustrate that every problem which can be solved using Ohm's Law can be readily solved using the new free form expression which replaces it. And problems which involve non-Ohm's-Law resistors and are difficult to solve using Ohm's Law science--they too are readily solved using the new free form expression.

The problems also illustrate that, in the new electrical science, problems involving nonlinear resistive devices are solved in *exactly the same way* as problems involving only proportional devices!

There are 4 problems. You will have no difficulty solving Problems 1 and 2 because they involve only proportional devices. But you will have considerable difficulty solving Problems 3 and 4 because they involve nonlinear devices, and you are not familiar with non-Ohm's-Law electrical science. If you solve Problems 3 and 4 using Ohm's Law science, you will have to do something you will not like. You will have to use resistance even though the resistance depends on the current--even though resistance has *no scientific value* in these problems!

CHAPTER 15

HOW TO USE THE NEW ELECTRICAL SCIENCE TO ANALYZE RESISTIVE CIRCUITS

It is quite likely that every person reading this book is able to analyze resistive circuits *provided* the circuits contain only devices which "obey" Ohm's Law--which behave in such a way that V/I is constant--which behave in a proportional way, and are commonly called "linear resistors".

It is also likely that not every person reading this book knows how to analyze resistive circuits which contain devices which do not "obey" Ohm's Law--which behave in such a way that V/I is not constant--which behave in a nonlinear way, and are commonly called "nonlinear resistors".

Ohm's Law science is not generally used to analyze circuits which contain nonlinear components, but analysis based on V=IR is possible if resistance is viewed as the *variable* ratio V/I. On the other hand, if resistance is viewed as the *constant* ratio V/I, then the resistance of nonlinear components is not defined, and analysis is *NOT* possible based on V=IR.

The problems in this chapter demonstrate how to use the new electrical science to analyze resistive circuits. They illustrate in a general way that:

- In the new electrical science, the solution of circuits which contain nonlinear components is *just as simple* as the solution of circuits which contain only proportional components.

- In the new electrical science, the solution of circuits which contain nonlinear components is *exactly the same* as the solution of circuits which contain only proportional components.

INTRODUCTION TO THE PROBLEMS

Problems (15-1) and (15-2) describe the new way analysis of circuits which contain only proportional components. Using today's electrical science, they would be solved with Ohm's Law science. Problems (15-3) and (15-4) illustrate the new way analysis of circuits which contain both proportional and nonlinear components. In today's electrical science, they would not normally be solved using Ohm's Law science.

If electrical resistance is viewed as the *constant* ratio V/I, then the resistance of nonlinear components is not defined, and it is not possible to solve Problems (15-3) and (15-4) using V=IR and resistance.

But if electrical resistance is viewed as the *variable* ratio V/I as in Halliday and Resnick (1978), it is possible to solve all 4 problems using V=IR and resistance. (Notice that electrical resistance is an analog of h *ONLY* when it is viewed as a variable ratio, since h is the variable ratio q/ΔT.)

Near the end of this chapter, the problem inputs are restated in terms of resistance viewed as a variable ratio. The reader is encouraged to solve the resistance versions of the problems using Ohm's Law and electrical resistance viewed as the variable ratio V/I.

You will have no difficulty solving the resistance versions of Problems (15-1) and (15-2). But you will probably have considerable difficulty solving the resistance versions of Problems (15-3) and (15-4) because they contain nonlinear components--because resistance has no scientific value in Problems (15-3) and (15-4).

In fact, you will probably give up before obtaining correct answers to these simple/difficult problems--simple using the new electrical science and behavior, quite difficult using today's Ohm's Law science and resistance viewed as a variable ratio.

PROBLEM (15-1)

What is the current in the circuit shown in Fig. (15-1)?

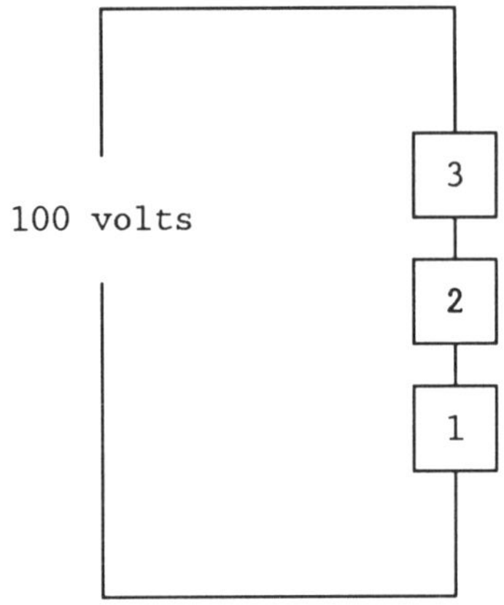

FIGURE (15-1)

RESISTIVE CIRCUIT IN PROBLEM (15-1)

INPUT TO PROBLEM (15-1)

The electrical behavior (ie the V{I} functions) of the components in Fig. (15-1) is given by:

$$V_1 = 3.0\ I_1 \tag{15-1}$$

$$V_2 = 4.5\ I_2 \tag{15-2}$$

$$V_3 = 6.2\ I_3 \tag{15-3}$$

ANALYSIS OF PROBLEM (15-1)

The voltage drops across series-connected components are additive, and therefore we may write

$$V_1 + V_2 + V_3 = 100 \tag{15-4}$$

Substituting Eqs. (15-1) to (15-3) into (15-4) gives Eq. (15-5). Since the currents are equal in series-connected devices, we may rewrite Eq. (15-5) in the form of (15-6).

$$3.0I_1 + 4.5I_2 + 6.2I_3 = 100 \qquad (15\text{-}5)$$

$$3.0I + 4.5I + 6.2I = 100 \qquad (15\text{-}6)$$

ANSWER TO PROBLEM (15-1)

The solution of Eq. (15-6) indicates that the current in Fig. (15-1) is 7.30 amps.

DISCUSSION OF PROBLEM (15-1)

It is important to note that the solution of this simple problem has nothing to do with the concept of electrical resistance. Nothing to do with the idea of resistance--nothing to do with the parameter R. The solution is based on the idea of electrical behavior--on V{I}.

The old way solution would be based on V, I, and R, even though the problem actually involves only V and I. R is not essential to the solution because R does not actually exist--it is an invention found only in the intellect--an invention which complicates and confuses the solution of problems by making it impossible to separate V and I.

Because V/I is independent of I in this simple problem, the solution is little affected by whether or not V and I are separated. That is why both new way and old way solutions are quite simple.

But when V/I varies with I, as in Problems (15-3) and (15-4), the difficulty of solution greatly depends on whether V and I are separated (as in the new way solution based on behavior) or are combined (as in the reader's solution of the restated problems at the end of this chapter).

PROBLEM (15-2)

What power is dissipated in Component 3 of Fig. (15-2)?

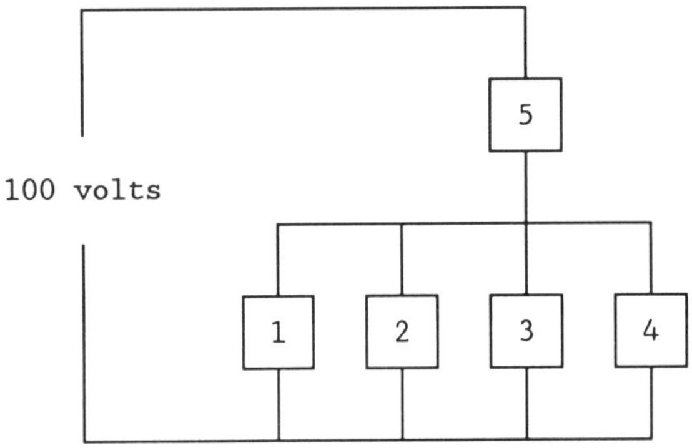

FIGURE (15-2)

RESISTIVE CIRCUIT IN PROBLEM (15-2)

INPUT TO PROBLEM (15-2)

The electrical behavior of the components in Fig. (15-2) is given by:

$$V_1 = 2.4\ I_1 \tag{15-7}$$

$$V_2 = 4.4\ I_2 \tag{15-8}$$

$$V_3 = 2.5\ I_3 \tag{15-9}$$

$$V_4 = 4.7\ I_4 \tag{15-10}$$

$$V_5 = 1.9\ I_5 \tag{15-11}$$

ANALYSIS OF PROBLEM (15-2)

Components 1 to 4 are equivalent to a single component described by

$$I_S = I_1 + I_2 + I_3 + I_4 \qquad (15\text{-}12)$$

$$V_S = V_1 = V_2 = V_3 = V_4 \qquad (15\text{-}13)$$

Substituting Eqs. (15-7) to (15-10) into (15-12) and using (15-13) gives

$$I_S = .417V_S + .227V_S + .400V_S + .213V_S \qquad (15\text{-}14)$$

This equivalent, single component is in series with Component 5, and therefore we may write

$$V_S + V_5 = 100 \qquad (15\text{-}15)$$

Combining Eqs. (15-11) and (15-15) gives

$$V_S + 1.9I_S = 100 \qquad (15\text{-}16)$$

Combining Eqs. (15-14) and (15-16) gives V_S=29.5, and therefore V_3=29.5. From this result and Eq. (15-9), we obtain I_3=11.8. Therefore, the power (P, watts) dissipated in component 3 is

$$P_3 = V_3I_3 = 29.5(11.8) = 348 \qquad (15\text{-}17)$$

ANSWER TO PROBLEM (15-2)

The power dissipated in Component 3 of Figure (15-2) is 348 watts.

PROBLEM (15-3)

What power is dissipated in Component 3 of Fig. (15-3)?

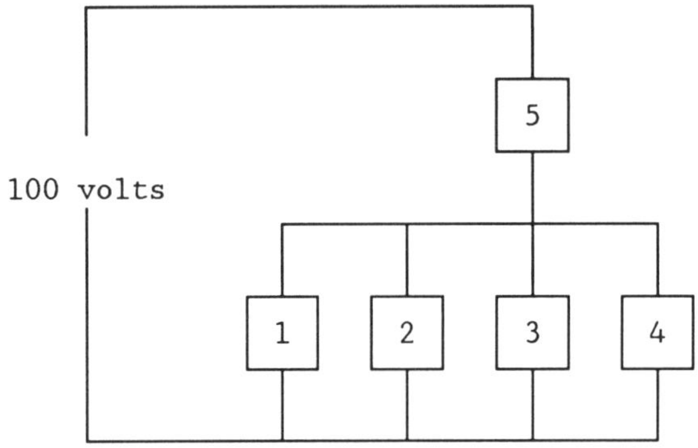

FIGURE (15-3)

RESISTIVE CIRCUIT IN PROBLEM (15-3)

INPUT TO PROBLEM (15-3)

The electrical behavior of the components in Fig. (15-3) is given by:

$$V_1 = 2.4I_1 \quad (15\text{-}18)$$

$$V_2 = 4.4I_2 \quad (15\text{-}19)$$

$$V_3 = 3.2(I_3)^{.5} \quad (15\text{-}20)$$

$$V_4 = 4.7I_4 \quad (15\text{-}21)$$

$$V_5 = 1.9I_5 \quad (15\text{-}22)$$

ANALYSIS OF PROBLEM (15-3)

Components 1 to 4 are equivalent to a single component described by

$$I_S = I_1 + I_2 + I_3 + I_4 \tag{15-23}$$

$$V_S = V_1 = V_2 = V_3 = V_4 \tag{15-24}$$

Substituting Eqs. (15-18) to (15-21) into (15-23) and using (15-24) gives

$$I_S = .417V_S + .227V_S + .0977(V_S)^2 + .213V_S \tag{15-25}$$

This equivalent, single component is in series with Component 5, and therefore we may write

$$V_S + V_5 = 100 \tag{15-26}$$

Combining Eqs. (15-22) and (15-26) gives

$$V_S + 1.9I_S = 100 \tag{15-27}$$

Combining Eqs. (15-25) and (15-27) gives V_S=17.2, and therefore V_3=17.2. From this result and Eq. (15-20), we obtain I_3=28.9. Therefore, the power dissipated in component 3 is

$$P_3 = V_3I_3 = 17.2(28.9) = 497 \tag{15-28}$$

ANSWER TO PROBLEM (15-3)

The power dissipated in Component 3 of Fig. (15-3) is 497 watts.

PROBLEM (15-4)

What is the current in the circuit shown in Fig. (15-4)?

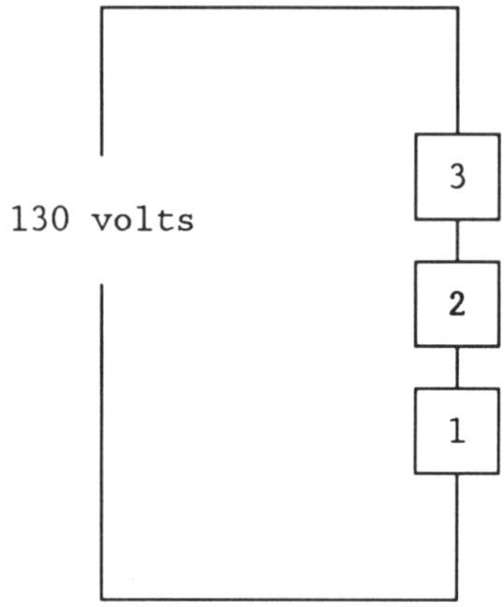

FIGURE (15-4)

CIRCUIT FOR PROBLEM (15-4)

INPUT FOR PROBLEM (15-4)

The electrical behavior of the components in Fig. (15-4) is given by:

$$V_1 = 4.1\ I_1 \qquad (15\text{-}18)$$

$$V_2 = 2.7\ I_2 \qquad (15\text{-}19)$$

$V_3\{I\}$: described in Fig. (15-5)

SOLUTION OF PROBLEM (15-4)

One way to solve this problem is to determine the overall electrical behavior of the circuit, $V_t\{I\}$, where the subscript "t" refers to total. Since voltage drops in series are additive, we may write

$$V_1 + V_2 + V_3 = V_t \qquad (15\text{-}21)$$

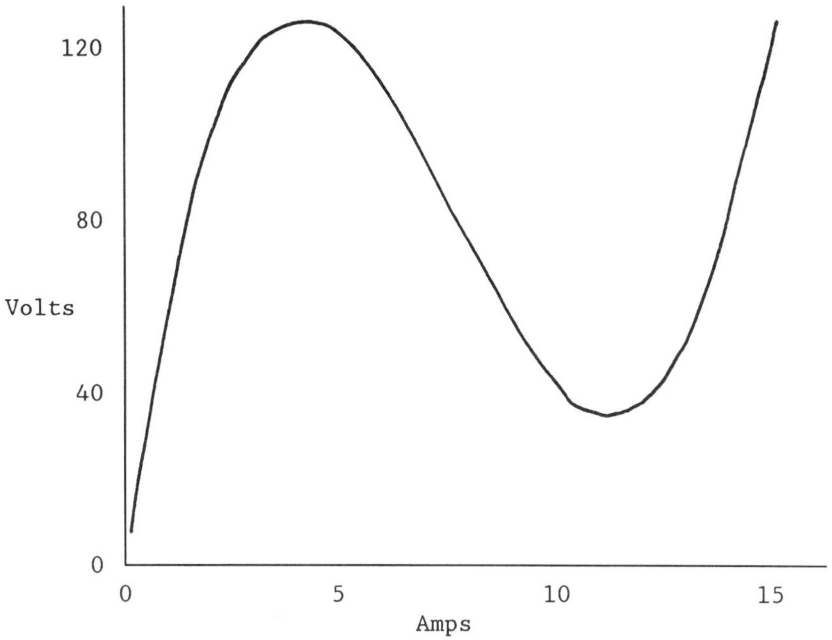

FIGURE (15-5) $V_3\{I\}$, Problem (15-4)

In order to determine $V_t\{I\}$, we must substitute the given $V\{I\}$ functions for the V's on the left side of Eq. (15-21). However, because the behavior of Component 3 is described graphically, it is simpler to determine $V_t\{I\}$ graphically than to convert Fig. (15-5) to an analytical expression.

The graphical determination of $V_t\{I\}$ is quite trivial:

- At selected I values, determine the voltage drop across each of the three components from the given information.
- Add the three voltage drops in order to obtain coordinates on $V_t\{I\}$.
- Plot the coordinates from the preceding step and draw a curve through the points. The curve is the desired function, $V_t\{I\}$. It is given in Fig. (15-6).
- In order to determine the current at a given value of total voltage drop, read $I\{V_t\}$ from Fig. (15-6).

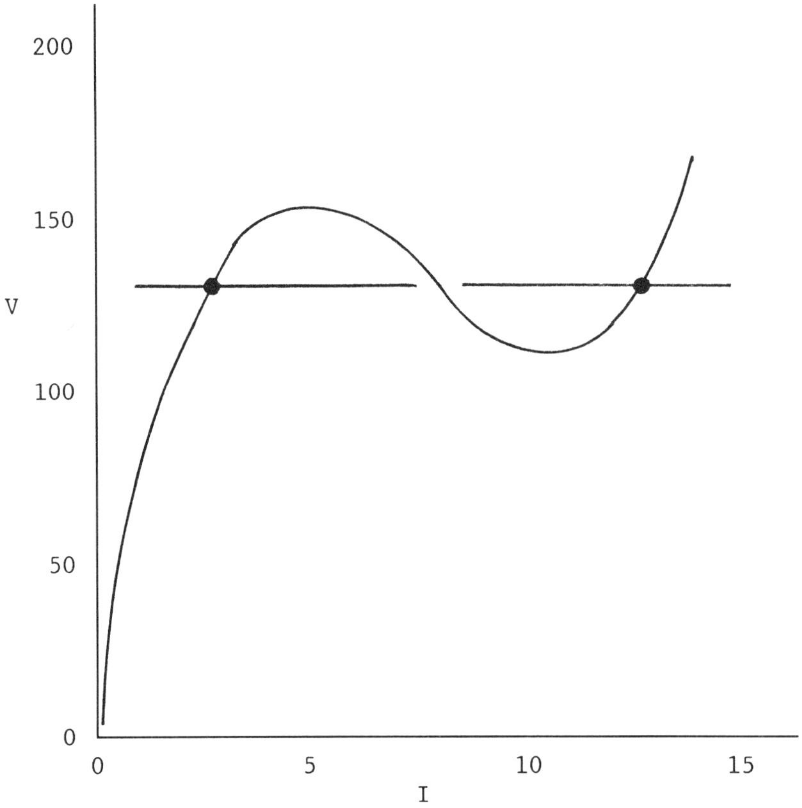

FIGURE (15-6) Solution of Problem (15-4)

- Since we are dealing with nonlinear behavior, we must be concerned about stability. Appraise the stability at potential operating conditions from the generic criterion

$$dV_{in}/dI <s \; dV_{out}/dI \qquad (15\text{-}22)$$

Notice that Criterion (15-22) is the electrical analog of the generic criterion for thermal stability.

ANSWER TO PROBLEM (15-4)

Inspection of Fig. (15-6) indicates that a supply voltage of 130 volts would result in a current of either 2.7 or 12.8 amps. The potential operating point at 8.0 amps does not satisfy Criterion (15-22) and therefore it is unstable; operation would shift to one of the two stable points. The given information does not uniquely define the solution to the problem. The prior operating history of the equipment would be required in order to determine a unique solution.

THE RESISTANCE VERSION OF PROBLEMS (15-1) TO (15-4)

The input to Problems (15-1) to (15-4) is restated below in terms of R viewed as the variable ratio V/I.

Problem (15-1) Input--Resistance Version:

R_1 = 3 ohms

R_2 = 4.5 ohms

R_3 = 6.2 ohms

Problem (15-2) input--Resistance Version:

R_1 = 2.4 ohms

R_2 = 4.4 ohms

R_3 = 2.5 ohms

R_4 = 4.7 ohms

R_5 = 1.9 ohms

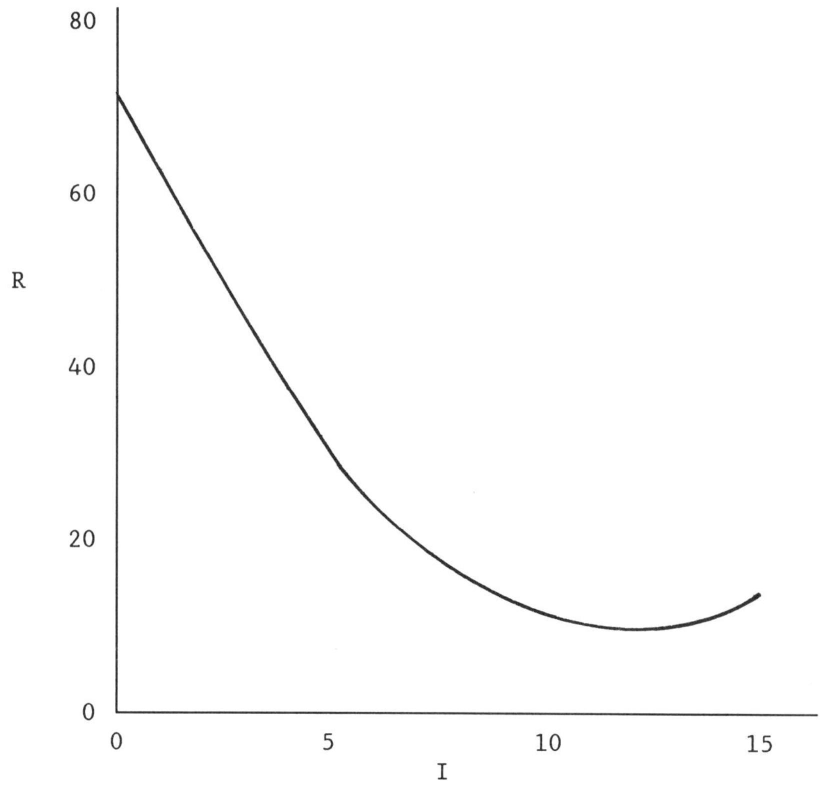

FIGURE (15-7) $R_3\{I\}$, Problem (15-4) *(old way)*

Problem (15-3) input--Resistance Version:

R_1 = 2.4 ohms R_2 = 4.4 ohms

$R_3 = 3.2\ (I_3)^{-.5}$ ohms

R_4 = 4.7 ohms R_5 = 1.9 ohms

Problem (15-4) input--Resistance Version:

R_1 = 4.1 ohms R_2 = 2.7 ohms

R_3 = described in Fig. (15-7).

The coordinates in Fig. (15-7) are (0,72), (2,55.5), (4,37.8), (6,25.3), (8,16.3), (10,11.1), (12,9.8), (14,11.9), (15,14.2).

THE SIGNIFICANCE OF PROBLEMS (15-1) TO (15-4)

The main purpose of Problems (15-1) to (15-4) is to demonstrate the new way solution of resistive circuits which include proportional and also nonlinear devices. These problems also demonstrate the following:

- The new way solution of resistive circuits is extremely simple, even if the circuits contain nonlinear devices.

- The new way solution of resistive circuits which contain nonlinear devices is *exactly the same* as the new way solution of circuits which contain only proportional devices.

- The new way solution of resistive circuits has nothing to do with the idea/concept/parameter of resistance.

- *Every* problem which can be solved with R can be solved better without it. If the problem involves only proportional devices (as in Problems (15-1) and (15-2)), there is little difference in the difficulty of solution. But as Problems (15-3) and (15-4) demonstrate, the new way solution is an order of magnitude simpler if the circuit includes nonlinear devices, particularly if the device is highly nonlinear and contains a region in which dV/dI is negative.

- Stability is a simple problem in the new electrical science. If the reader copes with the stability aspect of Problem (15-4), he will recognize that stability is extremely difficult to deal with in terms of resistance.

- The equation V=IR can be used to deal with nonlinear devices if resistance is defined to be the *variable* ratio V/I. The difficulty with doing this is that it is mathematically ineffective because it requires problems to be solved without separating V and I.

  That is the point intended to be made by the restated problems which the reader is asked to solve using Ohm's Law and resistance. The mathematical ineffectiveness of applying Ohm's Law science to problems involving nonlinear behavior is illustrated by the difficulty and tedium involved in the solution of Problems 3 and 4.

Using V=IR to deal with nonlinear devices is the electrical analog of using q=hΔT to deal with nonlinear heat transfer phenomena such as free convection or boiling. It can be done, but it involves a high degree of difficulty--difficulty which can be avoided by simply abandoning contrived parameters such as resistance and h--as in the new electrical science and the new thermal science.

In the new electrical science, we have no interest in any type of resistance, or in any Law based on resistance. We solve *ALL* resistive circuit problems in a logical and simple manner using V{I}, the new idea/concept of electrical behavior. And this new electrical science is closely analogous to the new thermal science.

In short, the problems are intended to illustrate that behavior is not merely an idea which results in a new thermal science. It is an idea which results in a new engineering science based on free form expressions and fundamental parameters--a new engineering science which has nothing to do with Laws and contrived parameters--a new engineering science which is logical and simple, and which has a great deal of commonality among its various branches.

CHAPTER 16

NEW VS TODAY'S OHM'S LAW SCIENCE

The immediate purpose of this chapter is to compare the analysis of resistive circuits using today's Ohm's Law science against the same analysis using the new science. But the greater purpose of this chapter is to demonstrate that

> *EVERY problem which can be solved with resistance and today's Ohm's Law science is better solved with behavior and the new electrical science.*

The comparison of today's methodology and the new methodology indicates that:

- The analysis of resistive circuits using resistance and V=IR is based on 9 equations.

- The analysis of resistive circuits using behavior and V=f{I} is based on only 5 equations.

- The 5 equations used in the new science are an integral part of today's Ohm's Law science.

- Resistance and the 9 equations deal effectively *ONLY* with proportional resistive circuits.

- Behavior and the 5 equations deal effectively with *ALL* resistive circuits--proportional, linear, and nonlinear.

The proper goal of science is logical simplicity. Because the new electrical science is much simpler and more logical than today's Ohm's Law science, the latter must give way.

THE EQUATIONS REQUIRED FOR THE ANALYSIS OF RESISTIVE CIRCUITS USING THE NEW ELECTRICAL SCIENCE

The analysis of resistive circuits using V{I} and the new electrical science requires the following 5 equations:

1. The voltage drops (V,volts) across series-connected components are additive:

$$V_{total} = \sum V_i \tag{16-1}$$

2. The voltage drops across parallel-connected devices are equal:

$$V_1 = V_2 = V_3 = \ldots \tag{16-2}$$

3. The currents (I,amps) through parallel-connected components are additive:

$$I_{total} = \sum I_i \tag{16-3}$$

4. The currents through series connected components are equal:

$$I_1 = I_2 = I_3 = \ldots \tag{16-4}$$

5. The power (P, watts) dissipated in an electrical component is given by the product of voltage drop and current:

$$P = V\ I \tag{16-5}$$

It is very important to note that these 5 equations and the new electrical science deal effectively with *ALL* purely resistive circuits including proportional, linear, and nonlinear.

(These 5 equations were sufficient to solve all the problems in the preceding chapter, and those problems included proportional components and highly nonlinear components.)

THE EQUATIONS REQUIRED FOR THE ANALYSIS OF RESISTIVE CIRCUITS USING TODAY'S OHM'S LAW SCIENCE

The analysis of resistive circuits using V=IR and today's Ohm's Law science requires 9 equations:

1. *Same as 1 above.*

2. *Same as 2 above.*

3. *Same as 3 above.*

4. *Same as 4 above.*

5. *Same as 5 above.*

6. Resistance (R, ohms) is defined to be the ratio of component voltage drop to component current:

$$R \equiv V/I \qquad (16\text{-}6)$$

7. The resistance of series-connected components is given by the sum of the component resistances:

$$R_{total} = \sum R_i \qquad (16\text{-}7)$$

8. The resistance of parallel-connected components is given by the reciprocal of the sum of the reciprocals of the individual components:

$$R_{total} = 1/\Sigma(1/R_i) \qquad (16\text{-}8)$$

9. The power dissipated in an electrical component is given by the product of current squared times resistance:

$$P = I^2 R \qquad (16\text{-}9)$$

It is important to note that these 9 equations and today's Ohm's Law science deal effectively *ONLY* with proportional devices. (If you made a serious effort to solve the resistance versions of Problems (15-3) and (15-4), recall that these 9 equations did *NOT* deal effectively with those problems because they involved nonlinear behavior.)

NEW SCIENCE VS TODAY'S SCIENCE

The analysis of resistive circuits (often called V{I} circuits in the new science) is much simpler using the new electrical science because the analysis is based on only 5 equations whereas Ohm's Law science requires 9 equations.

Moreover, the 5 equations and the new science deal effectively with *all* resistive circuits, whereas the 9 equations and today's Ohm's Law science deal effectively only with the simplest possible resistive circuits.

> *The 5 equations are not new. They have been an integral part of Ohm's Law science for almost 200 years.*

As far as equations for the new electrical science are concerned, the only requirement is to *FORGET* four old equations! In a very real sense, the reader already knows how to solve electrical circuits using the new science because the equations required by the new science are part of today's Ohm's Law science.

A CLOSE LOOK AT EQUATIONS (16-6) TO (16-9) AND RESISTANCE

Equations (16-6) to (16-9) have always been unnecessary in an absolute sense. Their seeming necessity resulted from the invention of "resistance"--from the intellectual artifice of viewing flow as the result of a driving force overcoming a resistance. The application of this generic scheme to electrical phenomena logically resulted in the invention of electrical "resistance".

Electrical resistance is an old concept--so old that it now seems to be real--to actually exist. But it does *NOT* exist! It "exists" only in the sense that phlogiston and caloric exist--only as a way to view Natural phenomena.

> *"Resistance" is simply a term we assign to V/I.*

When resistance became the accepted way to view electrical phenomena, it *THEN* became necessary to deal with V and I using resistance--using V/I. And that is precisely and uniquely the purpose of Eqs (16-6) to (16-9)--to make it possible to deal with V and I indirectly using V/I, rather than using V and I directly.

> *In the new electrical science, we choose to view electrical phenomena in terms of behavior rather than resistance, and therefore we have no use for resistance--no use for Eqs. (16-6) to (16-9).*

In Ohm's Law science, it was not consciously recognized that Eqs. (16-6) to (16-9) are unnecessary and redundant. They are unnecessary and redundant because they contain no information that is not also contained in (16-1) to (16-5).

The unnecessary/redundant nature of Eqs. (16-6) to (16-9) is seen by noting that there is no information in Eq. (16-6)--it merely states that R is a shorthand way to write V/I. Eqs. (16-7) to (16-9) are redundant because the information they contain is already contained in (16-1) to (16-5). This is readily demonstrated by substituting V/I for R in these equations, and then separating V and I.

For example, Eq. (16-7) is identical to Eq. (16-1). This is readily shown by substituting V/I for R in Eq. (16-7):

$$V_{total}/I_{total} = V_1/I_1 + V_2/I_2 + V_3/I_3 \ . \ . \ . \qquad (16\text{-}10)$$

Since the components are series-connected, Eq. (16-4) allows us to write

$$I_{total} = I_1 = I_2 = I_3 = \ . \ . \ . \qquad (16\text{-}11)$$

Combining Eqs. (16-10) and (16-11) and separating V and I gives

$$V_{total} = \sum V_i \qquad (16\text{-}12)$$

In the same way, it is readily shown that Eqs. (16-8) and (16-9) are identical to (16-3) and (16-5), and therefore they are redundant and unnecessary.

Eqs. (16-6) to (16-9) do nothing other than "translate" (16-1) to (16-5) from the language of V{I} to the language of V/I (alias R)--from the language of separated variables to the language of combined variables--from the language which is simple/clear/desirable to the language which is complex/foggy/undesirable.

CONCLUSION

The idea/concept/parameter of "resistance" has two harmful effects on electrical science:

- "Resistance" complicates the analysis of resistive circuits by making Eqs. (16-6) to (16-9) necessary, whereas they are unnecessary if resistance is not invented.
- "Resistance" is the reason today's Ohm's Law science cannot deal effectively with nonlinear circuits. As Maxwell correctly noted, resistance has "no scientific value" unless V/I is independent of I--unless all devices are proportional devices.

In the new electrical science, we abandon the idea of resistance and all resistance parameters. We invent the idea of electrical behavior. We combine this new idea with what is left of today's electrical science after resistance has been removed.

The new electrical science which results is an improvement over today's Ohm's Law science because the new science deals effectively with nonlinear devices, and Ohm's Law science does not.

CHAPTER 17

THE REPLACEMENTS FOR INDUCTANCE AND CAPACITANCE IN THE NEW ELECTRICAL SCIENCE

In the new electrical science, we abandon the parameters "inductance" (L, henries) and "capacitance" (C, farads), just as we abandon the parameter "resistance". We abandon them for the same reason we abandon resistance--because they are unnecessary and undesirable. And because they are unnecessary, we do not replace them with new parameters.

The new electrical science also abandons the *ideas* of inductance and capacitance. They are replaced by the same idea which replaces resistance--the new idea that

> *The simplest, most logical way to deal with electrical phenomena is in terms of V{I,t} and NOT in terms of parameters contrived from V, I, and t.*

Today's electrical science uses volts and amps and seconds and ohms and henries and farads and *TWO* completely different sciences to deal with electrical devices. The new electrical science uses volts and amps and seconds and *ONE* logical science to deal with electrical devices.

This chapter describes the new way to deal with inductive and capacitive devices. The description is conceptual in nature, and is presented by analogy to the development of the new way to deal with resistive devices.

This chapter completes the presentation of the new electrical science--for now. We shall return to this subject in a future, comprehensive book on the new electrical science.

INDUCTANCE IN TODAY'S ELECTRICAL SCIENCE

Inductors exhibit a wide variety of behavior, ranging from strictly proportional to nonlinear. The simplest purely inductive behavior is described by

$$V = L(dI/dt) \qquad (17\text{-}1)$$

For this simple type of behavior, inductance is defined by

> The inductance of an electrical device is defined to be the ratio of voltage difference to the derivative of electrical current with respect to time. In symbolic terms, L is defined to be the ratio of V to dI/dt where L is henries, V is volts, I is amps, and t is seconds.

Inductance has *no scientific value* when the value of V/(dI/dt) depends on the value of (dI/dt). In order to deal effectively with V and I when the value of V/(dI/dt) depends on the value of (dI/dt), it is necessary to invent other forms of inductance, and to use these other inductances in a way which little resembles the way in which the above inductance is used.

CAPACITANCE IN TODAY'S ELECTRICAL SCIENCE

Capacitors exhibit a wide variety of behavior, ranging from proportional to nonlinear. The simplest capacitors exhibit behavior described by

$$I = C(dV/dt) \qquad (17\text{-}2)$$

For this simple type of behavior, capacitance is defined by

> The capacitance of an electrical device is defined to be the ratio of the electric current to the derivative of voltage difference with respect to time. In symbolic terms, C is defined to be the ratio of I to the derivative dV/dt where C is farads, I is amps, V is volts, and t is seconds.

Capacitance has *no scientific value* when the value of I/(dV/dt) depends on the value of (dV/dt). In order to deal effectively with V and I when the value of I/(dV/dt) depends on the value of (dV/dt), it is necessary to invent other forms of capacitance, and to use these other capacitances in a way which little resembles the way in which the above capacitance is used.

CAPACITANCE AND INDUCTANCE IN THE NEW ELECTRICAL SCIENCE

Capacitance and inductance are closely analogous to electrical resistance. Notice that Eqs. (17-1) and (17-2) are closely analogous to Ohm's Law. Like Ohm's Law, they do *NOT* describe the behavior of capacitors or inductors in a general way.

Since the new electrical science has no use for resistance, it is not surprising that it has no use for capacitance or inductance, and that these parameters are not replaced in the new science. Nor is it surprising that capacitance and inductance are undesirable for the very same reasons that resistance is undesirable.

We could detail the development of the replacements for capacitance/inductance, but the development would be highly repetitious because it would echo the last few chapters.

The development of the replacement for inductance requires little more than a repetition of the last few chapters with L substituted for R, and dI/dt substituted for I. The development of the replacement for capacitance requires little more than a repetition of the last few chapters with C substituted for R, I substituted for V, and dV/dt substituted for I.

To avoid the repetition, we simply observe that inductance and capacitance are abandoned in the new electrical science for the same good reasons that we abandon resistance. The new view of inductance and capacitance is described in the following:

- The idea of inductance is abandoned. It is replaced by the idea of electrical behavior--the idea that the behavior of all electrical devices is best described and dealt with in terms of V{I,t} and *NOT* in terms of an inductance parameter contrived from V, I, and t.

- The parameter of inductance and its dimension (henries) are abandoned. They are not replaced because they are unnecessary/undesirable.

- The idea of capacitance is abandoned. It is replaced by the idea of electrical behavior--the idea that the behavior of all electrical devices is best described and dealt with in terms of V{I,t} and *NOT* in terms of a capacitance parameter contrived from V, I, and t.

- The parameter of capacitance and its dimension (farads) are abandoned. They are not replaced because they are unnecessary/undesirable.

- Eqs. (17-1) and (17-2) and other equations which define other types of inductance and capacitance are abandoned. They are all replaced by the generic expression

$$V = f\{I,t\} \qquad (17\text{-}3)$$

- The words inductance, capacitance, henries, and farads are abandoned. The words inductive and capacitive are retained with their current meanings.

CLOSING REMARKS

The preceding few chapters describe the conceptual basis of the new electrical science and demonstrate its application to the analysis of purely resistive devices. This chapter merely describes the new conceptual basis as it applies to inductive and capacitive devices. However, application in this area requires only a minor extension of the treatment in the preceding few chapters.

CHAPTER 18

THE OLD WAY AND THE NEW WAY TO GENERATE ENGINEERING SCIENCES

This chapter describes both the conventional method used to generate today's engineering sciences, and the new method I used to generate the thermal and electrical sciences described in this book.

The principal difference between the two methods is that the conventional method generates a science based on equations which describe behavior in a specific way, whereas the new method generates a science based on equations which describe behavior in a general way--a free form way. For example, Ohm's Law describes behavior in a specific way and is important in today's electrical science; the new equation, $V=f\{I\}$, describes behavior in a general way and is important in the new electrical science.

It seems odd to refer to the conscious generation of a particular science, or to present a detailed description of the methodology used in generating it. The generation of a science is generally viewed as the result of inspiration which cannot be described. But the generation of a science is a task like any other. The task is *not* accomplished by inspiration, and the method of accomplishing the task is readily examined.

In this chapter, we first examine the conventional methodology used to generate today's engineering sciences. Many were generated in the 19th century, so we must look there to find the source and application of conventional methodology. As is generally the case, a study of the old way is excellent preparation for the new way.

The primary purpose of describing the details of the new methodology is *not* to rationalize the new thermal and electrical sciences presented in Parts 1 and 2. The purpose is to describe how similar new sciences are to be generated in other branches of engineering.

The new engineering will not be truly utilized until all engineering sciences have been based on equations which describe behavior in a free form way, and all contrived parameters have been abandoned. Or more specifically, until all engineering sciences are based on equations which resemble Eqs. (18-1) to (18-7), and all "parameters" like k, h, R, E, f are abandoned.

$$q_{cond} = f\{dT/dx\} \qquad (18\text{-}1)$$

$$q_{conv} = f\{\Delta T\} \qquad (18\text{-}2)$$

$$V = f\{I\} \qquad (18\text{-}3)$$

$$\sigma = f\{\epsilon\} \qquad (18\text{-}4)$$

$$W = f\{\Delta P\} \qquad (18\text{-}5)$$

$$s = f\{t\} \qquad (18\text{-}6)$$

$$y = f\{x\} \qquad (18\text{-}7)$$

The new sciences are not simply different--they are also better. A great deal better because they make it possible to deal with all types of behavior in a simple and logical way. A scientific way.

TODAY'S VIEW OF NATURAL PHENOMENA AND LAWS

Today's view of Natural phenomena and the Laws is little different from the 19th century view that

> *The behavior of Natural phenomena is prescribed by simple Laws which dictate the relationships among Natural parameters.*

Discovering these Laws and formulating sciences which apply them to the solution of practical problems was and is considered the proper goal of science and the scientist.

In the very first sentence of his famous treatise, Fourier states:

> Primary causes are unknown to us; but are subject to simple and constant laws, which may be discovered by observation, the study of them being the object of natural philosophy (science).

In the first sentence of his equally famous treatise, Ohm expresses a similar view:

> The design of this Memoir is to deduce strictly from a few principles, obtained chiefly by experiment, the rationale of those electrical phenomena . . . which have been termed galvanic; its aim is attained if by means of it the variety of facts be presented as unity to the mind.

In the 19th century view, the Laws are not simple and constant in theory--the Laws are simple and constant *in fact*. Fourier does not suppose that the Laws are simple. The Laws *ARE* simple. Nature *IS* simple. She is not governed by complex Laws--nonlinear Laws. She is constant--She does not change the Law from day to day.

Fourier's purpose was to discover "the simple and constant laws" which govern heat flow phenomena. Ohm's purpose was to discover "the rationale" by which "the variety of facts (is) presented as unity to the mind". Fourier and Ohm discovered the Laws they sought, and the Laws were simple indeed. The Laws were proportional--the Laws were of the form y=mx--the simplest of all possible forms!

Fourier and Ohm discovered the simple and constant Laws they sought, and they developed them into true sciences which accurately and universally described the Natural phenomena which were important *in their time*.

THE CONVENTIONAL METHOD OF SCIENCE GENERATION

The conventional method of science generation is today (1989) little different from that utilized by Fourier and later by Ohm. Its essential elements are:

1. Deduce/induce the parameters of fundamental importance in the science. In other words, define and quantify the fundamental parameters (such as V and I), and determine how they are to be measured.

2. Deduce/induce which fundamental parameters are related to which others. For example, determine that V and I are somehow related--ie determine that

$$V = f\{I\} \tag{18-8}$$

3. Deduce/induce the behavior of the fundamental parameters relative to each other. For example, determine that the ratio V/I is constant--ie that V/I is independent of I. Or equally, determine that V is proportional to I.

4. Express the behavior of the fundamental parameters with equations. For example, express the result that V/I is constant by writing

$$V = m\ I \tag{18-9}$$

   where "m" is a pure constant in the equation and would be called a "constant coefficient" in mathematics.

5. Imagine a hypothetical view of the phenomenon in which the pure constants in the Laws are imagined to be Natural parameters which have physical character and dimensions. For example, imagine that electrical phenomena are the manifestation of a driving force V overcoming a "resistance" and resulting in a current I.

In this hypothetical view, we imagine that the pure constant "m" is the resistance parameter.

6. In accordance with the imagined view of the phenomenon, assign names, symbols, and dimensions to the pure constants in the Laws. For example, assign "m" the name "resistance", the symbol R, and the dimension ohms. Write the equation in accordance with the imagined view, and then call it a Law. For example, "Ohm's Law"

$$V = I R \qquad (18\text{-}10)$$

7. Develop the science based on the Law or Laws. In other words, generate the equations and ideas from which the Laws can be utilized to predict the magnitude of fundamental parameters in a variety of circumstances. For example, generate the equations and ideas from which Ohm's Law can be utilized to predict the magnitude of V and I throughout any purely resistive circuit.

8. Verify that the science is a true science by demonstrating close agreement between analytical predictions obtained from the science and empirical results obtained from a comprehensive set of experiments.

THE NEW VIEW OF NATURAL PHENOMENA AND THE LAWS

A true science must be universal. It must be able to deal with *ALL* phenomena within its domain in a simple and logical way. Therefore it must be based on expressions which describe Natural behavior in a *universal* way.

At their birth, the thermal and electrical sciences of Fourier and Ohm were true sciences--the proportional Laws on which they were based were universal descriptions of known behavior. But then it was discovered that Natural phenomena are *NOT* monolithic. Natural phenomena are infinitely varied and cannot be described in a universal way with Laws which prescribe behavior in a specific way.

Infinitely varied behavior can be described in a universal way only with an expression which is infinitely varied--such as y=f(x).

Ohm's Law is a simple Law which accurately and universally described the behavior of conductors which were important in the nineteenth century. But many of today's important electrical devices do not "obey" Ohm's Law.

Ohm's Law science is not a true science today because Ohm's Law does not describe the behavior of today's electrical devices in a universal way.

Fourier's so-called "Newton's Law of Cooling" is written in the form of a Law, but it is not a Law because it does not describe behavior. It merely defines h to be the variable ratio $q/\Delta T$. Since variable ratios have "no scientific value", "Newton's Law of Cooling" and h cannot be the bases for a true science.

Fourier's Law of Conduction is a simple Law which accurately and universally describes today's behavior. But what about tomorrow? Isn't it likely that tomorrow will bring materials which do not behave in the simple way described by this Law? Materials whose behavior will transform today's true science into tomorrow's untrue science? Wouldn't it be better to base conductive heat transfer on the analog of y=f(x) in order to increase the likelihood that it will be a lasting science?

In the new engineering, we insist that ALL engineering sciences be based on analogs of y=f(x) in order to deal effectively with ALL behavior known today, and ALL behavior to be discovered tomorrow.

THE NEW METHOD OF SCIENCE GENERATION

The method I used to generate the new thermal and electrical sciences is described below in generic terms:

1. Deduce/induce the parameters of fundamental importance, or obtain them from the existing science. In other words, define and quantify the fundamental parameters y and x, and describe how they are to be measured.

2. Deduce/induce which fundamental parameters are related to which others, or obtain this information from the existing science. For example, determine that y and x are somehow related--ie determine that

$$y = f\{x\} \qquad (18\text{-}11)$$

3. Develop the science based on the equation(s) determined in Step 2. In other words, generate the equations and ideas from which free form equations can be utilized to predict the magnitude of fundamental parameters in a variety of practical circumstances.

4. Verify that the science is a true science by demonstrating close agreement between analytical predictions obtained from the science and empirical results obtained from a comprehensive set of new or existing experiments.

NEW VS OLD METHODOLOGY

A comparison of new vs old methodology reveals that:

- Steps 1 and 2 in the new method are the same as Steps 1 and 2 in the old method.

- The new method omits Steps 3 to 6 in the old method. The purpose of these steps is to determine the Laws--the specific behavior from which the science is to be formulated. Since the new method is not based on specific behavior, these steps are unnecessary and are omitted.

- Step 3 in the new method is the same as Step 7 in the old method except that free form behavioral equations such as Eq. (18-11) replace the specific Laws which are the bases of today's engineering sciences.

- Step 4 in the new method is the same as Step 8 in the old method.

The differences between new and old methodology result from the new idea that engineering sciences can and should be based on equations which describe behavior in a free form way. The old idea was that engineering sciences should be based on equations which describe behavior in a specific way--the so-called "Laws" of today's engineering such as Ohm's Law, etc.

The end result of this new idea is that the new engineering sciences differ from the old sciences in the following respects:

- There are no behavioral Laws in the new sciences. No Ohm's Law, no Fourier's Law of Conduction, no Fourier's "Newton's Law of Cooling". All behavioral Laws are abandoned and replaced by free form behavioral equations. (Behavioral laws are distinguished from postulated laws such as the conservation laws. The latter laws are retained in the new sciences.)

- The new sciences contain no contrived parameters such as k, h, R, E, f, Nu, St. . . . They are not replaced because they are unnecessary and undesirable.

- The equations of the new sciences are not dimensionally consistent because dimensions are not assigned to pure constants in the new methodology. (The old methodology preserves dimensional consistency by arbitrarily assigning dimensions to pure constants.) Natural phenomena are not dimensionally consistent, and therefore the new equations more closely resemble Natural phenomena than their counterparts in the old engineering sciences.

THE COMPLETENESS OF THE NEW SCIENCES

The new methodology for generating sciences is very brief, but it results in sciences which are complete in every way.

In order to prove that a science is complete, it must be demonstrated that all problems within its domain can be solved without the addition of any other concept or parameter. This is proven by demonstrating that

- The y{x} functions can be determined.
- The y{x} functions can be dealt with when they have been determined.

It is obvious that the y{x} functions can be determined. We have merely to perform an experiment in which y and x vary, and then correlate the data in the form y{x}.

It is important to note that this correlation methodology is a vast departure from the methodology of today's engineering which usually correlates data in the form y/x--such as V/I (alias R), q/ΔT (alias h), and σ/ϵ (alias E).

It is also obvious that the y{x} functions can be dealt with because pure mathematics deals with y{x} functions in a very effective way. In order to deal with the analogs of y{x} in the new engineering sciences, we have merely to utilize the methods of pure mathematics.

It should be noted that today's engineering and the new engineering are equally empirical. The empiricism differs only in that the new engineering substitutes the form y{x} for the form (y/x){x} generally used in today's engineering sciences.

CONCLUDING REMARKS

The new method of generating engineering sciences is the direct result of the new idea that

> *Engineering sciences can and should be based on free form descriptions of Natural behavior. These descriptions should be written in terms of fundamental parameters, and should have NOTHING TO DO with contrived parameters.*

The new thermal and electrical sciences presented in this book are the end result of utilizing the new method of science generation. They demonstrate its application, its usefulness, and its general applicability.

The new sciences which result from the method are radically different from today's engineering sciences in both concept and application. Their particular advantage is that they make it possible to deal with all types of behavior in a simple and logical way.

Surely that is the proper goal of science.

CHAPTER 19

HOW FOURIER GENERATED HIS HEAT TRANSFER SCIENCE

Fourier and his contemporaries felt that Natural phenomena are governed by "simple and constant" Laws, and that Laws are the proper bases for the Natural sciences.

In accordance with this view, Fourier determined the fundamental parameters involved in heat transfer phenomena, and the Laws which dictate their behavior. The science he generated is a description of how these Laws are used to solve problems of a practical nature. Fourier's science has been somewhat revised, but it is still the essence of today's heat transfer science.

Our concern in this chapter is not with the heat transfer science Fourier generated, but rather with the methodology he used to generate it. The method Fourier used is still the conventional method--engineering sciences are still generated by discovering the Laws which describe (or seem to describe) Natural behavior, and then developing the ideas and equations which make it possible to apply these Laws to the solution of practical problems.

In this chapter, Fourier describes how he generated today's heat transfer science. In a later chapter, we will use the new method to generate the new heat transfer science. When we have used both methods, we will be better able to judge the relative merits of the two methods and their results.

FOURIER'S DETERMINATION OF THE FUNDAMENTAL PARAMETERS IN CONDUCTIVE HEAT FLOW

Fourier's search for the Law of Conductive Heat Flow did not begin with the knowledge that heat flux was somehow related to temperature gradient. His search began with the limited knowledge of conductive heat transfer which prevailed at the beginning of the 19th century:

- An understanding of the concept of heat.
- An understanding of the concept of temperature.
- The knowledge that the flow of heat through a solid body is generally proportional to temperature difference.

Fourier's contemporaries had little idea how or even that the flow of heat is influenced by the length and area of the heat flow path. Fourier tells us

> The quantity of heat which passes between (parallel planes) is in fact proportional to the difference (a-b) of the extreme temperatures, but it does not depend only on this difference: all other things being equal, it is less when the (planes are further apart), and in general it is proportional to (a-b)/e (where e is the distance between the parallel planes). . .
>
> We lay stress on this remark because the neglect of it has been the first obstacle to the establishment of the theory (of heat flow). If we did not (consider the dimensions of importance in heat flow), a forteriori, we should not be able to form the equations which express the movement of heat . . .
>
> It was necessary also to introduce into the calculation the dimensions of the (object being analyzed), in order that we might not regard, as general, consequences which observation had furnished in a particular case. Thus, it was discovered by experiment that a bar of iron, heated at one extremity, could not acquire, at a distance of six feet from the source, a temperature of one degree (Reamur); for to produce this effect, it would be necessary for the heat of the source to surpass considerably the point of fusion of iron;

(Fourier refers to a paper by Biot (1804). From his experiments with a bar of iron 6 feet long, Biot concluded that *ANY* bar of iron 6 feet long, heated at one end, would melt before the unheated end reached a temperature of even one degree. Fourier corrects Biot's erroneous conclusion:)

> but this result depends on the thickness of the (bar) employed. If it had been greater, the heat would have been propagated to a greater distance . . . We can always raise by one degree the temperature of one end of a bar of iron, by heating the solid at the other end; we need only give the radius of the base a sufficient length; which is, we may say, evident, and of which besides a proof will be found in the (quantitative) solution of the problem (given below). (p. 59)

(Notice that the above passages contain Steps 1 to 3 of the conventional method described in the preceding chapter.)

In these passages, Fourier states that his contemporaries could not solve even the simplest problems in conductive heat flow because they did not know that the heat flow rate depends on the area and length of the heat flow path.

Fourier states that heat flow rate is directly proportional to flow path area, and inversely proportional to flow path length. In other words, he was the first to determine that the fundamental parameters in conductive heat flow are heat flux and temperature gradient. He further states that they are related in a proportional way.

THE BASIS FOR FOURIER'S DEDUCTION OF THE HEAT FLOW LAWS

Fourier describes the "proposition" on which his deduction of the heat flow Laws was based:

> *If two molecules of the same body are extremely near, and are at unequal temperatures, that which is the most heated communicates directly to the other during one instant a certain quantity of heat; which quantity is proportional to the extremely small difference of the temperatures . . .*

> This proposition expresses a general and constant fact, which is sufficient to serve as the foundation of the mathematical theory. . .
>
> The general equations of the movement of heat, in the interior of solids of any dimensions, and at the surface of these bodies, are necessary consequences of the foregoing proposition. They are rigorously derived from it, as we proved in our first Memoirs in 1807, and we easily obtain these equations by means of lemmas, whose proof is not less exact than that of the elementary propositions of mechanics. (p. 456)

In short, Fourier deduced the Laws--"The general equations of the movement of heat"--from his "proposition" that heat flows from molecule to nearby molecule *in proportion* to their difference in temperature.

FOURIER'S DEDUCTION OF THE LAW OF CONDUCTIVE HEAT FLOW

Fourier deduced that the local heat flux within solid bodies is proportional to the local temperature gradient. But the Law he was seeking was not merely an expression of proportionality. The Law he was seeking was an *equation*.

In his treatise, Fourier deduces the macroscopic Law of Conductive Heat Flow from his microscopic "proposition" that heat flows from molecule to nearby molecule in proportion to their difference in temperature. He applies this proposition to the determination of temperature and heat flux distribution in "an infinite solid enclosed between two parallel and infinite planes". Figure (19-1) describes the nomenclature used by Fourier in the deduction of the Law of Conductive Heat Flow.

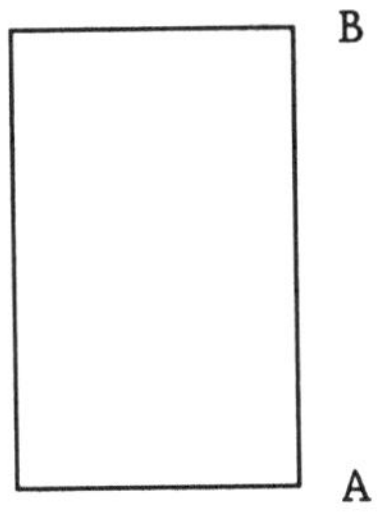

$b = T_B$

e = total height

z = height variable

$v = T(z)$

$a = T_A$

FIGURE (19-1) Fourier's nomenclature

Fourier concludes his deduction with:

> The value of F denotes the quantity of heat which, during the unit of time, passes across a unit of area of the surface taken on a section parallel to the base.
>
> Thus the thermometric state of a solid enclosed between two parallel infinite plane sides whose perpendicular distance is e, and which are maintained at fixed temperatures a and b, is represented by the two equations:
>
> $$v = a + (b - a)(z/e) \qquad (19\text{-}1)$$
>
> $$F = -K\,(dv/dz) \qquad (19\text{-}2)$$
>
> The first of these equations expresses the law according to which the temperatures decrease from the lower side to the opposite side, the second indicates the quantity of heat which, during a given time, crosses a definite part of a section parallel to the base.
>
> We have taken this coefficient K, which enters into the second equation, to be the measure of the specific conducibility of each substance; this number has very different values for different bodies.
>
> It represents, in general, the quantity of heat which, in a homogeneous solid formed of a given substance and enclosed between two infinite parallel planes, flows, during one minute, across a surface of one square metre taken on a section parallel to the extreme planes, supposing that these two planes are maintained, one at

> the temperature of boiling water, the other at the temperature of melting ice, and that all the intermediate planes have acquired and retain a permanent temperature.
>
> . . . We shall shew presently how to determine by observation the value K of the conducibility or conductibility in different substances. (p.52)

(Notice that these passages contain Steps 4 to 6 of the conventional method described in the preceding chapter.)

Eq. (19-2) expresses Fourier's deduction that heat flux is equal to a constant times the temperature gradient. Notice that he transforms this equation into a Law by transforming the constant coefficient into a parameter. The transformation is readily accomplished by arbitrarily assigning a name, a symbol, physical character, and dimensions to a *pure constant*!

Notice that if K is a pure constant, then Eq. (19-2) is merely an empirical and dimensionally inconsistent equation. But if K is a dimensioned parameter, then Eq. (19-2) is a dimensionally consistent Law! And of course Fourier had no difficulty choosing between these two alternatives.

FOURIER'S DEDUCTION OF THE LAW OF CONVECTIVE HEAT FLOW

Fourier deduced the Law of Convective Heat Flow from the "principle" that the flow of heat between nearby molecules is proportional to their difference in temperature. He states:

> The quantity of heat which bodies lose through their surface is subject to the same principle. If we denote by σ the area, finite or infinitely small, of the surface, all of whose points have the temperature v, and if "a" represents the temperature of the atmospheric air, the coefficient h being the measure of the external conducibility, we shall have $\sigma h(v-a)dt$ as the expression for the quantity of heat which this surface σ transmits to the air during the instant dt.

> When the two molecules, one of which transmits to the other a certain quantity of heat, belong to the same solid, the exact expression for the heat communicated is that which we have given (above); and since the molecules are extremely near, the difference of the temperatures is extremely small. It is not the same when heat passes from a solid body into a gaseous medium. But the experiments teach us that if the difference is a quantity sufficiently small, the heat transmitted is sensibly proportional to that difference, and that the number h may, in these first researches, be considered as having a constant value, proper to each state of the surface, but independent of the temperature. (p. 43)

> We have taken as the measure of the external conducibility of a solid body a coeffient h, which denotes the quantity of heat which would pass, in a definite time (a minute), from the surface of this body, into atmospheric air, supposing that the surface had a definite extent (a square metre), that the constant temperature of the body was 1, and that of the air 0, and that the heated surface was exposed to a current of air of a given invariable velocity. This value of h is determined by observation. The quantity of heat expressed by the coefficient is composed of two distinct parts which cannot be measured except by very exact experiments. One is the heat communicated by way of contact to the surrounding air: the other, much less than the first, is the radiant heat emitted. (p. 31)

(Notice that these passages contain Steps 4 to 6 of the conventional method described in the preceding chapter.)

Using modern symbols, the Law described above is

$$q = h \, \Delta T \qquad (19\text{-}3)$$

which Fourier intended to be a statement that convective heat flow is a proportional phenomenon, and that h is the proportionality constant--the constant ratio $q/\Delta T$.

It was necessary to decide whether h is a pure constant or a dimensioned parameter--whether Eq. (19-3) is an empirical and dimensionally inconsistent equation, or whether it is a dimensionally consistent Law. And again, Fourier had no difficulty making a choice.

FOURIER AND NONLINEAR HEAT TRANSFER BEHAVIOR

It is understandable that Fourier's heat transfer science does not address highly nonlinear heat transfer behavior, since two-phase heat transfer was not important at the beginning of the 19th century. But it is surprising that Fourier's science does not address moderately nonlinear behavior, since free convection was important at that time.

Fourier was aware that, during free convection, heat flux is not proportional to temperature difference. He states:

> When the heated body is placed in air which is maintained at a sensibly constant temperature, the heat communicated to the air makes the layer of the fluid nearest to the surface of the body lighter; this layer rises more quickly the more intensely it is heated, and is replaced by another mass of cool air. A current is thus established in the air whose direction is vertical, and whose velocity is greater as the temperature of the body is higher. For this reason if the body cooled itself gradually the velocity of the current would diminish with the temperature, and the law of cooling would not be exactly the same as if the body were exposed to a current of air at a constant velocity. (p. 30)

Fourier avoided dealing with the nonlinear heat flow which results from natural convection by the manner in which he defined h. Recall that his definition of h is based on the flow of heat from a body whose surface temperature is 1 degree above ambient, and which is "exposed to a current of air of a given invariable velocity". The end result is that Fourier's h is independent of ΔT *by definition*.

FOURIER'S DIMENSIONAL ANALYSIS

Fourier had no doubt that scientific equations must be "homogeneous" (dimensionally consistent). He states:

> In the analytical theory of heat, every equation expresses a necessary relation between the existing magnitudes x, t, T, C, h, k. . . . If we attribute to each quantity its own *exponent of dimension*, the equation will be homogeneous (dimensionally consistent), since every term will have the same total exponent. (pp. 128-129)

> On applying the preceding rule (which allows the dimensions of the coefficients to be determined) to the differential equations and their transformations, it will be found that they are homogeneous with respect to each kind of unit (dimension), and that the dimension of every angular or exponential quantity is nothing. If this were not the case, some error must have been committed in the analysis, or abridged expressions must have been introduced. (p. 130)

Fourier *assumes* that "every equation" is dimensionally consistent, but he gives no basis for this assumption, nor does he describe how it was verified. He presents this assumption in a manner which suggests that it is not open to question, even though it bears no resemblance to Natural phenomena.

Notice Fourier's observation that, if an equation which describes a Natural phenomenon is not dimensionally consistent, "some error must have been committed in the analysis". Equations which are dimensionally inconsistent are *wrong* equations! Therefore the constant coefficients in heat flow equations *must* have dimensions. Fourier is certain of that.

Fourier devised a method to determine the dimensions of the constant coefficients in the Laws--the dimensions of the pure constants which are arbitrarily assigned dimensions. Fourier's dimensional analysis assigns to the constants whatever dimensions make the Laws dimensionally consistent.

In other words, the method devised by Fourier is the same method used today. This method is based on pretending that the dimensions on the left side of equations equal those on the right side. This pretense is sufficient to determine the dimensions which must be assigned to the constant coefficients in order to preserve dimensional consistency.

It should be noted that the rationale of the method is faulty. The method involves making an assumption, and then making it come true in an artificial way. We assume dimensional consistency, and then we make it come true by arbitrarily assigning dimensions to pure constants. Which of course proves nothing and improves nothing.

We perform this exercise in "dimensional analysis" in order to preserve dimensional consistency. But why is it deemed necessary to do so? Why is it better to pretend that pure constants have dimensions than to recognize that they do not? Why is it better to pretend that the Laws are dimensionally consistent when in fact they are not?

The Laws only *appear* to be dimensionally consistent. We arbitrarily assign dimensions to pure constants in order to obtain the *appearance* of dimensional identity. But this exercise changes nothing. It is a truly useless exercise.

THE APPLICATION AND VALIDATION OF FOURIER'S HEAT TRANSFER SCIENCE

Fourier recognized that the Laws were valuable only insofar as they were useful in the solution of practical problems. He states:

> It is easy to judge how these (heat transfer) researches concern the physical sciences and civil economy, and what may be their influence on the progress of the arts which require the employment and distribution of heat. They have also a necessary connection with the system of the world, and their relations become known when we consider the grand phenomena which take place near the surface of the terrestrial globe. (p. 2)

> In general the theorems concerning the heating of air in closed spaces extend to a great variety of problems. It would be useful to revert to them when we wish to foresse and regulate temperature with precision, as in the case of green-houses, drying-houses, sheep-folds, work-shops, or in many civil establishments, such as hospitals, barracks, places of assembly. (p. 73)

For that reason, much of Fourier's treatise is devoted to a description and demonstration of how the Laws are used to solve practical problems. Fourier states:

> Such are the chief problems which have been discussed in this work; they are all directed to one object only, that is to establish clearly the mathematical principles of

> the theory of heat, and to keep up in this way with the progress of the useful arts, and of the study of nature. (p. 23)

Fourier recognized the need to demonstrate that analytical results obtained with his science closely accorded with empirical results. He states:

> I have deduced these laws from prolonged study and attentive comparison of the facts known up to this time: all these facts I have observed afresh in the course of several years with the most exact instruments that have hitherto been used. (p. 2)

> It still remained to compare the facts with theory. With this view, varied and exact experiments were undertaken, whose results were in conformity with those of analysis, and gave them an authority which one would have been disposed to refuse to them in a new matter which seemed subject to much uncertainty. (p. 22)

(Note that the passages quoted above contain Steps 7 and 8 in the conventional methodology described in the preceding chapter.)

CHAPTER 20

HOW OHM GENERATED HIS ELECTRICAL SCIENCE

Ohm shared the view held by Fourier that Natural phenomena behave in a manner prescribed by simple and constant Laws. Like Fourier, his objective was to discover the Laws and develop a science based on them. When Ohm's treatise was published, his science did not have the practical importance of Fourier's. Time would soon change that.

Ohm's treatise on electrical science was published in 1827, several years after the publication of Fourier's treatise on heat transfer. Ohm was familiar with Fourier's methods and results, and he noted the close parallel between the flow of heat and the flow of electricity.

In this chapter, Ohm describes the methodology he used to generate the electrical science which has lasted almost 200 years. It is essentially the same methodology used by Fourier. The one notable difference is that Ohm was not concerned with dimensional consistency. Ohm's version of what is now called Ohm's Law was Eq. (20-1) where L has the dimension of length.

$$I = V/L \qquad (20\text{-}1)$$

This equation is used throughout Ohm's treatise, and the fact that it is dimensionally inconsistent created not the slightest difficulty. Apparently, Ohm saw no reason to invent the requirement of dimensional consistency, and no reason to observe it.

OHM'S DETERMINATION OF THE FUNDAMENTAL PARAMETERS

Just as Fourier had determined the fundamental parameters in heat transfer science, it was necessary for Ohm to determine the fundamental parameters in electrical science. Lockwood (1891) states:

> Before (Ohm's) time, the quantitative circumstances of the electric current had been indicated, in a vague way, by the use of the terms "intensity" and "quantity", to which no accurately defined meaning was attached. Ohm's service consisted in introducing and defining the accurate notions--electromotive force, current strength, and resistance. He indicated the connection of these with experiment, and stated his famous law that the electromotive force divided by the resistance is equal to the strength of the current.

With regard to the fundamental parameter of electromotive force, Ohm (1827) states:

> (To measure the electroscopic force,) we take the body A of very small dimensions, so that when we bring it into contact with the part to be tested of any third body, it may from its smallness be regarded as a substitute for this part; then its electroscopic force, measured in the way described, will, when it happens to be different at the various places, make known the relative difference with regard to electricity between these places.
>
> The intention of the preceding explanation is to give a simple and determinate signification to the expression "electroscopic force" . . . (p. 108)

OHM'S DEDUCTION OF "OHM'S LAW"

Ohm describes his deduction of what has come to be called "Ohm's Law":

> *Memoir based on Three Laws*. Three laws . . . forms the basis of (this entire treatise) . . .
>
> The two latter are purely experimental laws; but the first from its nature is, in part at least, theoretical. With regard to this first law, I have started from the

> supposition that the communication of the electricity from one particle takes place directly only to the one next to it, so that no immediate (electrical current) from that particle to any other situate at a greater distance occurs. The magnitude of the (electrical current) between two adjacent particles under otherwise exactly similar circumstances, I have assumed as being proportional to the . . . electrical difference of the two particles . . . (p. 12)

Ohm deduced "Ohm's Law" from his "supposition" that electrical current takes place only between adjacent particles, and that its magnitude is *proportional* to the "electrical difference" of the adjacent particles.

Notice that Ohm deduced "Ohm's Law" in the same way Fourier deduced the heat flow Laws--from the principle/supposition that the magnitude of the flow from molecule to molecule is in proportion to their difference in potential.

OHM'S LAW ACCORDING TO OHM

Electrical "resistance" is a concept/parameter/idea which gives physical character to electrical phenomena, and it accurately describes Ohm's data. But it is not the only idea which does so.

There is at least one other idea which gives physical character to electrical phenomena, and it describes Ohm's data just as accurately as "resistance". This other idea is Ohm's idea--the idea of "reduced length"--the idea in Ohm's Law according to Ohm:

> *Laws of the Current*. . . (The second Law becomes)
>
> $$S = A/L \qquad (20\text{-}2)$$
>
> (S is current, A is electroscopic force, and L is reduced length) which is generally true, and already reveals the equality of the force of the current at all parts of the circuit; in words it may be expressed thus:

> *The force of the current in a galvanic circuit is directly as the sum of all the tensions, and inversely as the entire reduced length of the circuit . . .* (p. 50)

It will no doubt surprise many readers to learn that Ohm's version of Ohm's Law did not contain the resistance parameter. Throughout his treatise, Ohm used the parameter he called "reduced length". The parameter called "electrical resistance" did not become part of Ohm's Law until some time after the publication of his treatise.

Notice Ohm's observation that Eq. (20-2) "is generally true".

THE DIMENSIONAL INCONSISTENCY OF OHM'S LAW

The modern version of Ohm's Law is dimensionally consistent only if we consider that "ohms" is a pseudonym for volts per amp, just as "Hertz" is a pseudonym for cps. But Ohm's version of Ohm's Law contains "reduced length" in place of resistance, and it is not possible to consider that a length dimension is a pseudonym for volts/amp.

Ohm's expression of his Law was dimensionally inconsistent. For that reason, Fourier would have considered it a wrong equation--not a Law. Had Fourier deduced Ohm's Law, there can be little doubt that the Law would have been dimensionally consistent and would have contained the "electrical transfer coefficient" with the dimension volts/amp.

But a dimensionally consistent version of Ohm's Law would be no better than the dimensionally inconsistent version by Ohm. It would solve the same problems in the same way, and it would give the same results with the same degree of difficulty. A dimensionally consistent version would be no better, and it would really be no different.

In many of today's engineering sciences, we arbitrarily assign dimensions to pure constants in order to achieve "dimensional consistency". But the net effect of this exercise is *precisely zero*.

THE APPLICATION AND VALIDATION OF OHM'S LAW

Ohm not only deduced Ohm's Law, he also developed the equations and ideas necessary to apply the Law to the solution of problems. Most of his treatise is a description of these equations and ideas, and a demonstration of how they are used to describe and predict electrical phenomena.

Ohm recognized the need to validate his theory, and so he compared analytical results from his science with empirical results obtained by himself and others. Ohm states:

> A theory which asserts itself to be an enduring and fruitful one, must in all its consequences be in accordance with observation and experiment.
>
> It seems to me with respect to (Ohm's Law), this is sufficiently established, partly by the previous experiments of others, and partly by some performed by myself . . .

THE PARALLEL BETWEEN HEAT TRANSFER AND ELECTRICAL SCIENCE

Ohm recognized the close parallel between the heat transfer and electrical sciences. He states:

> The form and treatment of the differential equations which describe the propagation of electricity in bodies of any kind and form are so similar to those given for the propagation of heat by Fourier and Poisson, that even if there existed no other reasons, we might with perfect justice draw the conclusion that there exists an intimate connection between these natural phenomena . . .

CHAPTER 21

PREPARING TO GENERATE THE NEW HEAT TRANSFER SCIENCE

My experience with heat transfer began in 1953 when I performed a simple heat transfer experiment in an undergraduate laboratory. During the next decade, my work involved heat transfer design/analysis/experiment, but always with heat transfer behavior which was essentially proportional. Throughout this period, I used the old heat transfer and its coefficients, and found them effective and satisfactory.

My first confrontation with the reality of highly nonlinear heat transfer came in 1963. Early that year, I obtained the position of project engineer in charge of a 300 KW boiling liquid metal test facility, a position I held for 6 months.

Heat transfer to a boiling liquid metal is a highly nonlinear phenomenon, and in short time, I became dissatisfied with heat transfer coefficients. That dissatisfaction was the first step in generating the new heat transfer science.

In the decade prior to 1963, I had developed a well defined attitude toward the purpose of experiment and the proper treatment of data. My attitude toward them may not be unique, but it bears little resemblance to the attitude reflected in the archival literature on today's heat transfer science.

In this chapter, I describe my attitude toward experiment and data in order to prepare the reader for the next chapter, and in the hope that the reader will come to share the attitude which made it necessary and simple for me to generate the new heat transfer science.

WHAT IS DATA?

Before we continue, we must determine what is and what is not data. Let us agree on the following:

> "Raw data" refers to the values of *individual* parameters measured in an experiment.
>
> "Reduced data" refers to the values of *individual* parameters obtained by reducing raw data.
>
> "Data" refers to both raw and reduced data.

The key word in the above is *individual*. Only individual parameters are data. If we combine two or more parameters into a group ratio or a group product, the value of the group is *NOT* a datum--the value of the group is a result. Whether the group is dimensional or dimensionless has no bearing on whether its value is a datum or a result. No group value is a datum. There are no exceptions.

For example, if you divide your height by your weight, the result is a group parameter with the dimensions ft/lb. But even though the value of this group parameter is obtained from data, the value of the group is not a datum--it is a result.

Let us call your height-to-weight ratio your "height coefficient", and let us assign it the symbol H. Also, let us invent "Newton's Law of Growing" where L is height and W is weight:

$$L = H\,W \qquad (21\text{-}1)$$

(It is misleading to write Eq. (21-1) as an equation, since it is not a phenomenological description of the relationship between L and W. It is merely "the defining equation for H" and should be written in the form of a definition, since that is exactly what it is. Notice that this "equation" is dimensionally consistent if we arbitrarily assign to H whatever dimensions are in the group L/W.)

Now reflect that, even though you have had a considerable interest in your height and weight for years,

- You have NEVER computed your height/weight ratio (alias height coefficient).
- No application form has EVER asked for your height/weight ratio (alias height coefficient), although most ask for your height and weight *separately*.
- You have NO IDEA how or whether your height/weight ratio (alias height coefficient) has increased or decreased since you were born.

There is a very good reason why you know nothing about your height coefficient. The reason is that your height coefficient contains so little information about you.

Suppose I tell you that my height coefficient is 0.031 ft/lb. Am I tall? Am I short? Am I fat? Am I thin?

You have no idea because the height coefficient is such a worthless statistic. It is worthless because it is a group parameter--it is a variable ratio--the very thing Maxwell observed has "no scientific value".

Suppose I tell you that my height coefficient increased from .028 to .031 ft/lb in the last 18 months. Did my height increase or decrease? Did my weight increase or decrease? And of course you cannot tell. My height coefficient contains so little information that you simply have no idea what is happening to my height or weight.

The height coefficient would have value only if it remained constant from birth until death. If this were the case, each person's height coefficient could be used to determine his weight from his height, or conversely.

In summary, H has some value if it is a constant ratio, but it has essentially no value if it is a variable ratio. *And it is a variable ratio.*

What information would be better than the height coefficient? The data! The data is *infinitely* better. If you know the value of my height and weight, you know something about me. If you know the value of my height coefficient, you know essentially nothing about me.

And of course the point is that when two or more data values are combined in a group "parameter", the information in the data is largely destroyed. There is a very simple way to avoid destroying the information in data:

- *Never* allow a group of parameters to attain a character of its own.

- *Never* assign a symbol to a group of parameters.

- If a group of parameters is truly important, refer to the group in terms of its individual parameters.

- *Always* report the values of individual parameters.

THE PURPOSE OF EXPERIMENT--NEW WAY

The following is a concise statement of my attitude toward the purpose of experiment:

> *The purpose of experiment is to determine the functionality among the parameters measured in the experiment--the functionality among the data.*

I doubt that anyone would seriously disagree with this attitude. "Obviously" we perform experiments to determine how the measured parameters are related to each other. What other reason could there be?

Although the above attitude is obviously rational, it is not the attitude reflected in the archival heat transfer literature. If it were, one would expect to find two things which are sorely missing from today's heat transfer literature:

- Tables which present data.

- Graphs which describe data.

For example, suppose we perform a fluid flow experiment to determine the hydraulic characteristics of 70 F water flowing through 1" d. pipes, and we are interested only in the "fully developed" regime. We install static pressure taps at intervals in the pipe, we perform the experiment by stepwise varying the flow, and we record the flow and pressure distribution at each step.

When the experiment is completed, the question arises:

> *Should we describe what we found?*
> *Or should we describe something else?*

Should we describe how the measured flow rate and the measured pressure gradient were related to each other? Should we describe the function $(\Delta P/\Delta x)\{W\}$ which describes what was actually found in the experiment? Or should we describe something else?

My attitude permits only one answer:

> *Describe the data. Describe what was actually found. Describe the coordinates $(\Delta P,x,W)$ in DIGITAL form. On LINEAR paper, plot the function $(\Delta P/\Delta x)\{W\}$.*

When this task of cardinal importance has been completed, then we may consider generalizing the results. But it should be recognized that generalization is based on theory and assumption, and therefore generalized results lack the permanent value possessed by a description of the actual results--the data.

THE PURPOSE OF EXPERIMENT--TODAY'S WAY

The attitude reflected in today's heat transfer literature is described in the following concise statement:

> *Don't describe the data.* Describe something else. Combine the parameters into dimensionless groups and report the values of the groups. That way, it will be impossible to tell what was actually done or found in the experiment.

And if this seems incredible, it is readily verified by glancing through any current heat transfer or fluid flow journal. Try to find data--try to find a figure which has an axis labeled heat flux or temperature difference or flow rate or pressure gradient. It will be a futile attempt. You will find only Nusselt number or Reynolds number or friction factor or Grashof number or Stanton number or Rayleigh number or modified Rayleigh number or . . . .

Today's methods have little use for data. One does not analyze data--data is too specific--too parochial. It is far better to "nondimensionalize" the data--to "expand" the data so that it no longer describes a specific experiment, but describes the entire universe instead! It doesn't matter that this method of generalization largely destroys the information which was the sole reason for performing the experiment. It doesn't matter that this method of generalization replaces information of permanent value with information of temporary value.

For example, if we transform $\Delta P\{W\}$ data to the dimensionless groups called "friction factor" and "Reynolds number", we can describe the hydraulic characteristics of *ALL diameters and ALL incompressible fluids at ALL temperatures.* From ONE experiment using ONE pipe and ONE fluid at ONE temperature!

The current methods stress dimensionless groups and largely ignore data. Before the experiment, "the important dimensionless groups" are determined by analysis or from a literature review. Then, after the data is obtained, it is transformed to the important dimensionless groups. *Before it is analyzed!!* The end result of this methodology is that the data is *NEVER* analyzed! Only dimensionless groups are analyzed.

If the above experiment were reported in a journal article, the results would probably be presented in graphical and analytical form. The graph would be on log log paper, and it would describe the experimental results by showing how the dimensionless group parameter called "friction factor" is related to the dimensionless group parameter called "Reynolds number". The analytical form would almost certainly be a power law which relates "f" to "Re". And the article would contain no data in digital form or any other form.

From the above information, it would be virtually or identically impossible to tell what was done or found in the experiment.

In short, today's old heat transfer attitude toward the purpose of experiment bears little resemblance to mine. As reflected in the literature, today's attitude is:

> The purpose of experiment is to determine the relationships among deduced dimensionless groups, based on the assumption/speculation/hope/wild guess that Nature prefers power laws.

THE DESIGN OF ONE PHASE, FORCED CONVECTION HEAT TRANSFER COEFFICIENT EXPERIMENTS AND THE DATA REDUCTION

In the decade preceding 1963, I several times performed experiments which included the "measurement" of forced convection heat transfer coefficients to one phase (liquid) water at high pressure. The test facility was a pressurized "loop" which contained an electrically heated test section.

The conventional method of performing this type of test in this type of facility is:

- Set the water flow rate at the desired value.
- Set the heat flux at a value which results in a boundary layer ΔT large enough to permit accurate measurement. Record the data at steady-state.
- Repeat the above at different values of water flow rate.
- Reduce the data to heat transfer coefficients by dividing the measured value of q by the measured value of ΔT. Or rather, reduce the data to "Nusselt numbers" by multiplying q times a "characteristic length" and then dividing by the boundary layer ΔT and the "heat transfer conductivity" of the fluid. Also reduce the data to "Reynolds numbers" by multiplying . . . .

This conventional method is not the method I used because it does not accord with my view that the purpose of experiment is to determine the relationship among the measured parameters. The experiment involves the measured parameters q and ΔT, but the above method does nothing to determine the relationship between q and ΔT. The above method *assumes* that the relationship is proportional, and it fails to test this assumption.

Since we can readily determine the relationship between q and ΔT from the experiment, my view is that we should do so, even though we have strong a priori reason to suppose that the relationship is proportional. Moreover, the conventional method described above requires that the data be reduced to h values by dividing q by ΔT. I do not use this conventional method because I devised a much better way to obtain h results--a way that requires a different experiment design, and a great deal more data.

My better method requires that heat transfer coefficient experiments be performed in the following way:

- Set the water flow rate at the desired value.
- Stepwise set the heat flux at six or seven values including zero. At each value, record data.
- Repeat the above at different levels of flow rate.
- For each flow rate, prepare a q vs ΔT plot on linear paper.
- Draw a straight line through the data on each plot IF the data suggest a straight line which passes near (0,0).
- Determine the h values by noting that the slope of each straight line equals h at that flow rate.
- (Instead of drawing lines and measuring slopes, the data can also be reduced to h values by linear regression, provided the data suggest straight lines.)

This method worked quite well, and always indicated that q and ΔT were essentially proportional because the q vs ΔT data points indicated a straight line on linear graph paper, and the straight line passed near (0,0).

It is important to note that, if q and ΔT had not been proportional, the lack of proportionality *would* have been detected by my experiment design (because the data would not have indicated a straight line which passed near 0,0). On the other hand, the lack of proportionality *would NOT* have been detected by the conventional design because it is based on assumed proportionality, and nothing is done to test this questionable assumption.

THE DESIGN OF BOILING, FORCED CONVECTION HEAT TRANSFER COEFFICIENT EXPERIMENTS, AND THE DATA REDUCTION

Forced convection h experiments which involve saturated boiling must also consider the effect of vapor quality (χ, no dimensions) on h. In other words, vapor quality must be varied in the test matrix, and the impact of vapor quality on h must be described in the h correlation which is the object of the experiment.

If it is recognized that boiling h's often depend on ΔT, the experiment design will reflect that ΔT is a correlating parameter, and the test matrix will include several levels of ΔT as well as levels of vapor quality, flow rate, and pressure for each configuration to be tested. If this is not recognized, ΔT will not be included in the test matrix, and the test results will be compromised.

Data reduction involves dividing the q data by the ΔT data to obtain h results, and then correlating the results in the form h vs χ, with fluid flow rate and pressure as correlating parameters, and perhaps also ΔT. If ΔT is not included as a correlating parameter, the data reduction will necessarily assume that $q/\Delta T$ is a constant ratio because the value of h will be measured at only one value of ΔT for each test condition.

If the experiment design does not include levels of ΔT in the test matrix, the experimenter has no idea whether he is dealing with proportional or nonlinear behavior. In the likely event the boiling heat transfer is nonlinear, the nonlinearity sends no signal to the experimenter, and his method seems to work satisfactorily except that there is a good deal of scatter in the results.

The impact of boiling on my method is that vapor quality is added to the test matrix, but the data is still analyzed in the form q vs ΔT. The impact on the data reduction is that vapor quality is simply one more correlating parameter. My method still requires that h be determined from the slope of a straight line through data points plotted on linear coordinates IF the data points describe a straight line which passes near (0,0).

If the data points do NOT describe a straight line which passes near (0,0), then $q/\Delta T$ is a variable ratio. And since we agree with Maxwell that variable ratios "have no scientific value", we *refuse* to describe the empirical results in terms of $q/\Delta T$ (alias h).

If we refuse to use h, we will be confronted with the following questions:

How will we reduce/correlate data without h?

How will we design/analyze equipment without h?

The formulation of these questions was the first step in the generation of the new heat transfer science and the new engineering.

CHAPTER 22

HOW THE NEW HEAT TRANSFER SCIENCE WAS GENERATED

The generation of the new heat transfer science began in 1963, during the six month period I was in charge of a boiling liquid metal test facility. The facility was devoted to a NASA program to obtain design correlations for a forced convection, liquid metal boiler to be used in the space program.

The program included obtaining the data from which a heat transfer coefficient correlation could be determined. I therefore designed an experiment to yield the data required by my method of reducing heat transfer data to h values. (The method involves obtaining h values from the slopes of straight lines drawn through q vs ΔT data points drawn on linear paper. See Chapter 21.)

In the course of the data reduction, I was confronted with the following dilemma:

> Draw a straight line through a graph of q vs ΔT data points which indicate a high degree of curvature.
>
> *OR* abandon h.

Since the first alternative was obviously unacceptable, I chose the second one. This raised two questions:

- *How can heat transfer data be reduced and correlated without h?*
- *How can heat transfer equipment be designed and analyzed without h?*

This chapter describes how I generated the new heat transfer science which is the generic answer to these questions and others like them.

THE TEST FACILITY

The facility test boiler was a double pipe, counterflow heat exchanger with liquid metal fluids in both the primary and secondary systems. The nominal operating temperatures were 2200 F (primary) and 1800 F (secondary).

The heat flux distribution in the boiler was determined from the measured temperature profile on the primary side of the boiler and the measured primary system flow rate. The ΔT across the boiling interface was obtained from thermocouples imbedded in the inner pipe of the boiler, and from the measured temperature and pressure of the secondary fluid. The vapor quality (χ, no dimensions) profile in the boiler was determined from the measured secondary system flow rate and integration of the measured heat flux.

The facility is described in detail in quarterly progress reports on NASA Contract NAS 3-2528. The reports were prepared by the General Electric Co., and are entitled "Alkali Metals Boiling and Condensing Investigations".

HOW SHOULD A STRAIGHT LINE BE DRAWN THROUGH CURVED DATA?

The usual method of reducing heat transfer data to heat transfer coefficients is to divide q data by ΔT data. This conventional method bears little resemblance to my method of determining h from the slope of straight lines drawn through q vs ΔT data points plotted on linear paper. (My method requires a great deal more data than the usual method because it requires a group of data points for the determination of a single h value. The method is described in detail in Chapter 21.)

Using my method, a separate plot was prepared for each test condition--ie each operating point at which the local values of χ, W, and P were within a narrow range. Figure (22-1) qualitatively describes a typical data plot at a typical test condition.

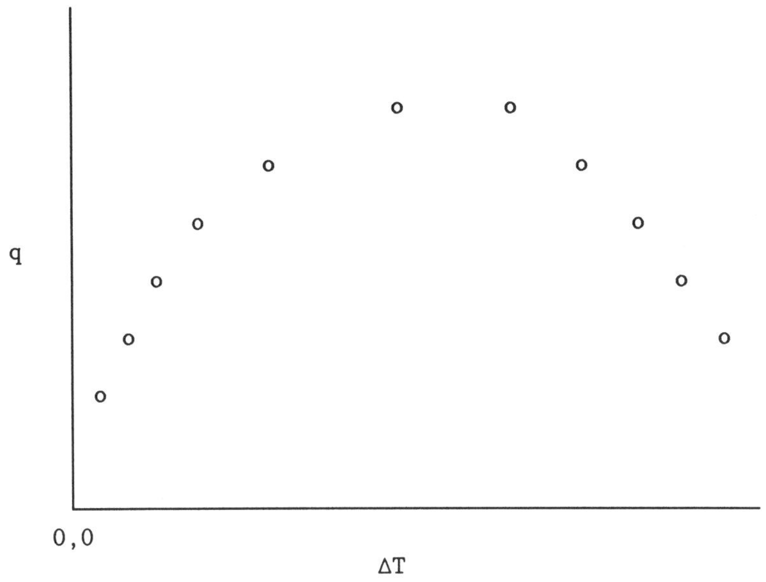

FIGURE (22-1) Qualitative illustration of q{ΔT} data points at fixed values of χ, W, P.

There is a problem with Fig. (22-1). The problem is that the method is based on the expectation that the data points will describe a straight line through the origin, whereas the data points in Fig. (22-1) obviously suggest something far different.

In other words, the method expects the figure to describe the same value of h at each data point, since all the data points were obtained at the same values of χ, W, and P. But Fig. (22-1) obviously contains many different h values. In fact, since no two points lie on the same straight line through the origin, each data point describes a different value of h.

(Recall that, at any point on a linear plot of q{ΔT}, the value of h equals the slope of the straight line between the origin and the data point.)

Before proceeding, it was necessary to answer the question:

> *How should a straight line be drawn in Fig. (22-1) in order to describe the heat transfer behavior at this test condition?*

And of course the answer is that there is *NO WAY* to draw such a line because *NO* straight line describes the heat transfer behavior at the test condition in Fig. (22-1).

(Notice that, if the data in Fig. (22-1) were reduced to h values in the usual way by calculating the value of $q/\Delta T$ at each data point, no difficulty would be encountered in reducing the data, but there would be considerable scatter in the $(q/\Delta T)\{\chi\}$ results--ie the $h\{\chi\}$ results.

Also notice that conventional experiment design requires only one data point at each test condition. If only one value of h were measured at each test condition, there would be no data scatter at any test condition. In other words, there would be no scatter within test conditions because there would be no degrees of freedom within test conditions.)

If we correlate the data by drawing straight lines which ignore the curvature in the data, we will obtain an h correlation which is obviously unreliable because it does not agree with the data. If this correlation is used to design the production boiler, its design will be less than optimum. The poor design will result because the correlation implicitly assumes that h is uniquely determined by local conditions other than heat flux.

But if we refuse to correlate the curved data with straight lines, we must be prepared to answer the questions:

- How will we reduce the data?
- How will we correlate the data?
- How will we design and analyze the production boiler which is the ultimate goal of the program?

THE NEW WAY TO DESCRIBE AND CORRELATE HEAT TRANSFER DATA

The conventional way to describe and correlate forced convection, boiling heat transfer data for a specific geometry is in the form $h\{\chi\}$, with flow rate and pressure as correlating parameters. This method implicitly assumes that $q/\Delta T$ is a constant ratio--ie assumes that the value of "h" is independent of q (or equally ΔT).

If it is recognized that $q/\Delta T$ is a variable ratio, the variability can be accounted for in one of two ways:

- Correlate the data in the usual form $(q/\Delta T)\{\chi\}$, and treat ΔT or q as an additional correlating parameter. (Of course $(q/\Delta T)\{\chi\}$ is commonly written $h\{\chi\}$.)

- Correlate the data in the form suggested by the data points in Fig. (22-1). The data points are values of q and ΔT, and the axes are q and ΔT. Both strongly suggest that the data be correlated in the form $q\{\Delta T\}$ with χ, W, and P as correlating parameters.

The first method is analogous to the manner in which free convection data are usually correlated in the old heat transfer. The outward form of these correlations is Nu{Ra} which is an opaque form of $h\{\Delta T\}$ which is an opaque form of $(q/\Delta T)\{\Delta T\}$. This last form is obviously undesirable because the variables q and ΔT are combined rather than separated.

Similarly, if we correlate boiling data in the form $h\{\chi\}$ with ΔT as a correlating parameter, this form is obviously undesirable if we recognize that it is actually the form $q/\Delta T\{\chi\}$ with ΔT as a correlating parameter.

On the other hand, if we correlate the data in the form $q\{\Delta T\}$ with χ as a correlating parameter, this form is obviously desirable because q and ΔT are separate. We therefore conclude that this form is most desirable. It is in fact the form used to correlate forced convection boiling data in the new heat transfer.

THE NEW WAY TO ANALYZE HEAT TRANSFER EQUIPMENT

The key steps in analyzing heat transfer equipment are:

- Calculate the local heat flux.
- Integrate the local heat flux over the heat transfer surface in order to determine the total heat flow rate.

In equipment in which heat is transferred from one fluid to another, the local heat flux is calculated by first calculating local values of h and k, then using these values to calculate the "overall heat transfer coefficient" (U, Btu/hr ft^2 F) from the expression

$$U = \frac{1}{1/h_1 + 1/h_2 + t/k} \tag{22-3}$$

The local heat flux and the total heat flow rate are then determined from

$$q_{local} = U_{local} \Delta T_{local} \tag{22-4}$$

$$Q_{total} = \int q_{local} \, dA \tag{22-5}$$

In order to determine how heat transfer equipment can be analyzed without h, it was necessary to determine how q could be calculated from q{ΔT} correlations instead of h correlations. In other words, it was necessary to find a substitute for Eqs. (22-3) and (22-4).

Since Eq. (22-3) is the key to the calculation of local heat flux in the old heat transfer, it seemed more than likely that some transformation of this equation would be a similar key in the new heat transfer.

In order to accomplish this transformation, it seemed necessary to answer the question:

What exactly are h and k?

(The answer to this question is now "obvious", but it was certainly not obvious to me in 1963. In fact, answering this question was the single most difficult and time consuming step in generating the new heat transfer science.)

As we have discussed before, my answer to this question is

- h is the ratio $q/\Delta T$. h is a variable ratio.
- k is the ratio $q/(\Delta T/\Delta x)$ as $\Delta x \rightarrow 0$.
  Today (1989), k is considered a constant ratio.

(If it seems incredible that I did not know the answer to this question in 1963, I remind the reader that the answer to this question was not within the knowledge envelope in 1963. The proof of this observation lies in the fact that h was universally the basis for heat transfer texts and heat transfer literature. This situation could have prevailed only if h were viewed as a parameter of Nature rather than a variable ratio "of no scientific value".)

When I recognized exactly what h and k are, it became "obvious" how to perform the required transformation of Eq. (22-3). Simply substitute $q/\Delta T$ for h, and $q/(\Delta T/\Delta x)$ for k, then separate q and ΔT. The result is:

$$\Delta T_{total} = \Delta T_1 + \Delta T_2 + \Delta T_{wall} \qquad (22\text{-}6)$$

Eq. (22-6) is the result of transforming Eq. (22-3) from an h expression to a $q\{\Delta T\}$ expression--from the old heat transfer to the new heat transfer.

Given Eq. (22-6), the problem is to determine how a value of q can be obtained from this expression which involves only ΔT's. The solution to this problem lies in recognizing that each ΔT on the right side of Eq. (22-6) is a function of q, and therefore "ΔT" implies "$\Delta T\{q\}$". In other words, Eq. (22-6) is an implicit form of :

$$\Delta T_{total} = \Delta T_1\{q_1\} + \Delta T_2\{q_2\} + \Delta T_{wall}\{q_{wall}\} \qquad (22\text{-}7)$$

In order to determine the local heat flux from Eq. (22-7), the functions on the right hand side must be given or determined. These functions are then substituted in Eq. (22-7), and the equation rearranged to determine (22-8), the desired expression for local heat flux:

$$q_{local} = f\{\Delta T\}_{total,local} \tag{22-8}$$

This new way of solving for local heat flux is demonstrated in the example problems in Chapter 3. It is simple to use, even when dealing with highly nonlinear phenomena.

The integration of local heat flux in order to determine total heat flow rate is the same in the new and old heat transfers, and therefore there is no new/old difference in integration to be discussed.

To summarize:

- In the old heat transfer, equipment analysis is based on calculating local heat flux by substituting h and k expressions into equations like (22-3) and (22-4), then determining total heat flow rate by integration.
- In the new heat transfer, equipment analysis is based on calculating local heat flux by combining $q\{\Delta T\}$ and $q\{dT/dx\}$ expressions with equations like (22-7), then determining total heat flow rate by integration.

THE NEW FOUNDATION FOR CONDUCTIVE HEAT TRANSFER BEHAVIOR

Today (1989), Fourier's equation

$$q = -k(dT/dx) \tag{22-9}$$

is a phenomenological description of conductive heat transfer behavior, and k is a constant ratio parameter which is universally applicable. Therefore Eq. (22-9) and k provide a sound basis for a conductive heat transfer science, and they could be included in the new heat transfer science.

But suppose that in the next few years, materials are developed which do not "obey" Eq. (22-9)--materials which exhibit a highly nonlinear relationship between q and dT/dx. If/when that happens, Eq. (22-9) and k will no longer provide a sound basis for a conductive heat transfer science, and they will have to be abandoned. Just as h and Fourier's "Newton's Law of Cooling" must now be abandoned.

To me, it seems ill advised to retain Eq. (22-9) and k until they must be abandoned. It seems far better to abandon them now, at the same time we are abandoning h and Fourier's "Newton's Law of Cooling". For that reason, I did not include Eq. (22-9) and k in the new heat transfer science, even though it is not absolutely necessary to abandon them now.

I determined the replacements for Eq. (22-9) and k by analogy to their convective counterparts. The new equation which replaces Eq. (22-9) is

$$q = f\{dT/dx\} \qquad (22\text{-}10)$$

There is no k parameter in this equation or in any part of the new heat transfer, just as there is no h parameter.

In the new heat transfer science, the conductive heat transfer behavior of materials which exhibit proportional behavior is described by proportional equations which resemble Eq. (22-9) except that they do *NOT* contain the dimensioned parameter k, and they are *NOT* dimensionally consistent. They contain a pure number in place of k, and they are dimensionally INconsistent.

The conductive heat transfer behavior of tomorrow's non-linear materials will be described by specific equations in the form of Eq. (22-10). They will contain the dimensioned parameters q and dT/dx, and various pure numbers. Every equation will be dimensionally INconsistent.

NEW VS OLD THEORY

Fourier entitled his treatise "The Analytical Theory of Heat". The theory was his deduction of the Laws/concepts

$$q = -k(dT/dx) \tag{22-11}$$

$$q = h\,\Delta T \tag{22-12}$$

The deduction of these Laws/concepts occupies a small fraction of Fourier's treatise. Mostly it is a description and demonstration of how to apply these Laws and concepts to the quantitative solution of practical problems.

In the same vein, the new theory is the deduction that heat transfer science can and should be based on the following new equations/concepts:

$$q_{cond} = f\{dT/dx\} \tag{22-13}$$

$$q_{conv} = f\{\Delta T\} \tag{22-14}$$

$$q_{radn} = f\{T\} \tag{22-15}$$

This deduction occupies a small fraction of this book. Mostly it is a description and demonstration of how to apply these equations and concepts to the solution of practical problems.

CHAPTER 23

BEHAVIOR CORRELATIONS

Behavior correlations are the bases for equipment design and analysis in the new engineering. They describe Natural phenomena in terms of fundamental parameters which appear in *explicit* form. (Fundamental parameters are found in Nature and can not be contrived from other parameters).

In the new engineering, there are no resistance/conductance/coefficient/factor/modulus parameters such as R, k, h, Nu, Re, f, E. These parameters are contrived, and we insist on parameters which are not contrived--fundamental parameters such as q, T, W, P. These are the parameters which truly concern us--the parameters which explicitly appear in behavior correlations--the only parameters used to design and analyze equipment in the new engineering.

In the next several chapters, we review the texts and literature on the old heat transfer to find heat transfer correlations which we translate into the language of the new heat transfer. However, we do not merely translate the correlations from old to new heat tansfer. We also examine the data which underlie them in order to verify that the correlations we generate agree with the reported data.

We find that several correlations little resemble the data they purport to describe, and we attribute the discrepancies to lack of rigor in today's old heat transfer method of generating correlations. Where we find conflict between old data and old correlation, we base the new correlations on the old data. The end result is that several of the new correlations differ from the corresponding old correlations in substance as well as form.

RESISTANCE/CONDUCTANCE/FACTOR/COEFFICIENT/MODULUS PARAMETERS IN BEHAVIOR CORRELATIONS

There are *NO* resistance/conductance/factor/coefficient/ modulus parameters in behavior correlations--no R, k, f, h, Nu, Re, St, E. These parameters are contrived, and contrived parameters play no part in behavior correlations or in the new engineering. They are replaced by the fundamental parameters from which they are contrived.

BEHAVIOR CORRELATIONS AND DIMENSIONAL ANALYSIS

The correlations of today's old heat transfer are usually written in terms of dimensionless groups, and these groups are largely the result of an intellectual process called "dimensional analysis". This process is independent of experiment, and is based on the assumption/speculation that Natural phenomena are "dimensionally consistent".

The end result of this old methodology is that h correlations are more the result of deduction than the result of experiment. They reflect experiment only in that data is required to establish the values of constants which appear in the deduced correlations, since the values of constants cannot be determined by dimensional analysis.

Eq. (23-1) is a typical result of dimensional analysis. It is the generic form of today's old heat transfer correlations for one phase, forced convection heat transfer:

$$Nu = a\ Re^b\ Pr^c \qquad (23\text{-}1)$$

The dimensionless groups Nu, Re, and Pr are obtained a priori by dimensional analysis, and experiment is required only to establish the values of a, b, c.

In the old heat transfer, a priori deduction is viewed as a scientific and desirable way to generate correlations. In the new heat transfer, a priori deduction is viewed as an UNscientific and UNdesirable way to generate correlations. Results obtained by a priori deduction are viewed with skepticism unless the deduced functionality of *EACH* fundamental parameter in the correlation is demonstrated. This demonstration is seldom (if ever) found in today's old heat transfer literature.

In Chapter 25, we discuss film cooling and the film cooling correlation:

$$\eta = f\{x/ms\} \qquad (23\text{-}2)$$

This generic and dimensionless correlation is deduced a priori, and it is today widely used in many high-tech universities and industries. We will use Eq. (23-2) to demonstrate that a priori deduction is an undesirable way to generate correlations because it leads to correlations which bear *no resemblance* to Natural behavior. In particular, we will find that Eq. (23-2):

- Seems to state that η depends on m when in fact it states that η *in no way* depends on m.

- Seems to state that η depends on s when in fact it states that η *in no way* depends on s.

- Oftentimes results in power laws which state that η equals infinity at x equal to zero, even though η cannot be greater than unity!

- Oftentimes leads to the conclusion there are two regimes in film cooling, even if the data indicates only one regime.

This film cooling example validates the conclusion that

> *A priori deduction is unreliable. It should be replaced by the scientific method of Galileo.*

BEHAVIOR CORRELATIONS AND DIMENSIONLESS PARAMETERS

There are *NO* dimensionless parameters in the behavior correlations of the new engineering. No Nusselt number, no Reynolds number, no Stanton number. They are not permitted because they are contrived parameters, and only fundamental parameters *in explicit form* are permitted in behavior correlations.

For example, in today's old engineering, a heat transfer correlation in the form of Eq. (23-1) would be acceptable. But this form is not acceptable in the new heat transfer. In order to transform Eq. (23-1) to an acceptable form, Nu, Re, and Pr must be replaced by the parameters from which they are contrived, the heat transfer properties must be converted to the form F{T}, and q and ΔT must be separated. The transformation of Eq. (23-1) to a heat transfer behavior correlation results in

$$q = a(G)^b F\{T\} (x)^{b-1} \Delta T \qquad (23\text{-}3)$$

Notice that Eq. (23-3) is a fluid specific correlation because F{T} describes the behavior of a particular fluid. The fluid general form of Eq. (23-1) is:

$$q = a(k_0/x)(xG/\mu)^b (C\mu/k_0)^c \Delta T \qquad (23\text{-}4)$$

Generalized correlations such as Eq. (23-4) are not considered behavior correlations because fluid temperature is necessarily implicit. Generalized correlations such as Eq. (23-4) are seldom used in the new heat transfer. They are largely replaced by fluid specific correlations which describe the behavior of a single fluid and in which fluid temperature (rather than fluid properties) appears in explicit form.

FLUID SPECIFIC BEHAVIOR CORRELATIONS

In today's old engineering, fluid specific correlations are seldom used, even in high-tech industries which have used only one heat transfer fluid for decades. Generalized correlations such as Eq. (23-1) are the usual basis for experiment and for equipment design and analysis.

If an experiment is performed in which air is the heat transfer medium, the data could be used to generate a correlation which specifically describes the measured behavior of heat transfer to air. In the old engineering, this is usually not done, even if the correlation is intended to be used exclusively for air. The usual procedure is to first reduce the data to Nu, Re, and Pr, then correlate the results in the generalized form of Eq. (23-1).

In the new engineering, fluid specific correlations are used almost exclusively. Their particular advantages are:

- They are simpler to use because all parameters which have design importance appear explicitly, and fluid properties do not appear at all.

- They are likely to be more accurate than generalized correlations because they are not based on the assumptions involved in generalizing.

- They are insensitive to revisions in fluid property tables.

BEHAVIOR CORRELATIONS AND POWER LAWS

Behavior correlations sometimes contain power laws, but *ONLY* if power law behavior is indicated by the data.

In the old heat transfer, data are often correlated with power laws without seriously questioning whether the data indicate power law behavior. The usual methodology is to assume power law behavior, plot the data on log log paper, draw the best straight line through the data, measure the slope of the line, and equate the slope of the line to the exponent in the assumed power law.

As we shall discuss in the chapters on nucleate boiling and transition boiling, this conventional methodology does not constitute a rigorous demonstration that the data indicate power law behavior. A simple and more rigorous demonstration of power law behavior would be to plot the data on linear paper, draw the best curve through the data, then show that this best curve resembles a power law.

When we apply this more rigorous treatment to data reported in the literature on the old heat transfer, we will find that highly nonlinear power law correlations oftentimes resulted from data which was essentially linear.

THE TRANSPARENCY OF BEHAVIOR CORRELATIONS

Behavior correlations are "transparent" in that they clearly display the relationship among the parameters which truly concern us--q, T, W, P. In this regard, the heat transfer correlations of the old heat transfer are opaque because they contain only parameters which do not truly concern us--Nu, Re, St.

We do not design Nusselt number exchangers. We design heat exchangers. We do not design friction factor equipment. We design fluid flow equipment. We do not design resistance transformers. We design voltage transformers. We are truly concerned with heat and flow rate and volts--these are the parameters on which we should focus our attention--the parameters we should use to design and analyze equipment--the parameters we do in fact use to design and analyze equipment in the new engineering.

We have no use for Nu or Re or h or k or f or E or any of the myriad contrived parameters of the old engineering. And since we have no use for them, we simply abandon them.

CHAPTER 24

CORRELATION METHODOLOGY--OLD AND NEW

In this chapter, we discuss the generic methodology used to investigate and correlate Natural phenomena in today's old heat transfer and in the new heat transfer. These methodologies bear little resemblance to each other.

Today's old methodology is described in a general way by the following:

- Parametric functionality is determined by the deductive process called "dimensional analysis".
- Correlations are written in terms of dimensionless group parameters such as Nu, Re, St.
- Correlations are power laws. This correlation form is selected by assumption.
- Correlations are "dimensionally consistent" because it is considered an obvious requirement.

The new methodology is essentially the opposite of the old. It is described in a general way by the following:

- Parametric functionality is determined by induction--by the "scientific method" of Galileo.
- Correlations are written in terms of fundamental parameters such as T, q, V.
- Correlations of all forms are considered and allowed. The correlation form is selected by induction.
- Correlations are *NOT* dimensionally consistent because "dimensional consistency" is an intellectual exercise which bears no resemblance to Natural phenomena.

THE CORRELATION METHODOLOGY OF TODAY'S OLD HEAT TRANSFER

In today's old heat transfer, correlations are largely obtained *a priori*, usually by the deductive process known as "dimensional analysis". This process is based on the following implicit assumptions, and no effort is generally made to verify them.

- Natural phenomena behave in a way which is dimensionally consistent. In other words, the dimensions on the left side of correlations must equal the dimensions on the right side.
- Natural phenomena are generally described by power laws.
- The effect of individual parameters is separable. In other words, if y=wz, the function y{w} does not depend on the value of z.

A priori deduction usually leads to dimensionless correlations in the form

$$DG1 = a(DG2)^b \qquad (24\text{-}1)$$

where DG refers to a dimensionless group identified by deduction. There are usually two or three DG's in a correlation. Since it is generally not possible to deduce the values of a and b by dimensional analysis, their values are determined by experiment.

Experiment design is relatively unimportant in today's old heat transfer because functionality is not determined by experiment--it is determined by deduction. Experiment design requires merely that the dimensionless groups be varied over a wide range in whatever manner is most convenient.

For example, if DG2 contains velocity and length parameters, it is not necessary to vary both of them in the experiment. Variation in DG2 results if either velocity or length is varied, and therefore the required variation in DG2 can be achieved by varying either velocity or length, whichever is more convenient.

The data are used merely to determine the values of a and b in Eq. (24-1). This is accomplished by plotting DG1 and DG2 on log log graph paper, drawing the best straight line through the points, and determining the values of a and b from the straight line.

The goodness of the correlation is demonstrated by showing that there is good overall agreement between Eq. (24-1) and the experimentally determined DG1{DG2}. This demonstration is accomplished by plotting Eq. (24-1) and the measured (DG1,DG2) points on the same log log graph of DG1 vs DG2. It is assumed that good overall agreement also demonstrates that the correlation accurately describes the functionality of the individual parameters in DG1 and DG2.

If the plotted points indicate pronounced curvature on the log log graph, it is concluded that the test matrix included several regimes. In this case, the curve is approximated by several straight line segments, and each segment is considered a "regime". A different power law correlation is required for each regime.

Because the old heat transfer correlations are in generalized form, they describe behavior in a universal way, even if the underlying experiment is of limited scope. For example, a generalized correlation describes the behavior of many fluids, all geometries, all velocities, and all temperatures. Even if the experiment on which it is based involved only one fluid, one geometry, one velocity, and one temperature.

THE NEW CORRELATION METHODOLOGY

In the new heat transfer, parametric functionality is largely determined by experiment and induction, and therefore the design of the experiment is extremely important. Experiments are designed so that *each fundamental parameter* in the experiment is varied over a wide range. This is necessary in order that the experiment will provide the data needed to establish functionality with respect to each varied parameter.

(As we noted above, the old heat transfer requires only that each *group* in the correlation be varied. Since the number of groups is less than the number of fundamental parameters, the old heat transfer does not require that each fundamental parameter in the correlation be varied.

In the old heat transfer, this is viewed as an important advantage. In the new heat transfer, this is viewed as an important shortcoming because it results in determining parametric functionality by deduction rather than experiment. Deduction has never been a reliable way to determine parametric functionality.)

Experiment design is simplified by assuming separability--ie by assuming that the fundamental parameters in the experiment act independently. Since this assumption is questionable, it is verified by experiment. The experiment design calls for this verification.

Correlations are induced by plotting the reduced *data* on *LINEAR* paper, and then formulating an expression which resembles the curve drawn through the plotted data points. (Log log paper is NOT used for this purpose.) All types of correlation forms are considered and allowed, including power laws.

The experimental correlation is written only in terms of the fundamental parameters which were varied over a wide range in the experiment. If one or more important parameters was not varied in the experiment, those parameters do not appear in the correlation which describes the experimental results.

Parameters not varied in the experiment may appear in an extension of the experimental correlation based in part on deduction or similarity to other experiments. Unvaried parameters in the extended correlation are flagged in order to point out to the potential user that their functionality was not determined by experiment and therefore should be considered unreliable.

Correlation goodness is demonstrated in two ways:

- By showing that the correlation agrees with the data in

an overall way--ie by comparing predicted result with measured result.

- By showing that the correlation accurately describes the functionality of each fundamental parameter in the experiment. This requires plotting correlation results and measured results against each fundamental parameter with the other parameters held at fixed values. These plots are also on linear paper.

NO group parameters are permitted in new way correlations. Group parameters simply do not exist in the new heat transfer. They are replaced by the fundamental parameters from which they are contrived in the old heat transfer.

THE NEXT SEVERAL CHAPTERS

In the next several chapters, our purpose is to compare the new and the old correlation methodologies. Our goal is to appraise their relative ability to result in correlations which accurately describe Natural phenomena.

We review the old methodology by studying its application to a more or less typical experiment described in the literature on the old heat transfer. We then use the new methodology to consider how the same experiment would be designed, how the data would be reduced, and how the results would be correlated if the experiment were based on the methodology of the new heat transfer.

The next several chapters demonstrate that the old methodology is undesirable because, in the example cited:

- It failed to identify the parameters of primary importance, and concluded that parameters of only secondary importance were parameters of primary importance.

- It identified several regimes when in fact the data indicated a single regime.

- It resulted in a correlation which did not resemble the underlying data.

CHAPTER 25

THE OLD HEAT TRANSFER VIEW OF FILM COOLING

Film cooling has been used in gas turbine and rocket applications for almost 50 years. It cools the surface of a hot flow path by the injection of cooler fluid into the boundary layer. The intent is that the cooler fluid will remain in the boundary layer for some distance, and will maintain the surface temperature below the temperature of the flow path fluid.

Although film cooling has several important applications, our purpose in considering it here has little to do with its application. Our purpose is to use film cooling as a vehicle to compare the old and new methodology of investigating and correlating heat transfer phenomena.

In this chapter, we discuss the current, widely held view of film cooling--the view which is the basis for the design and analysis of high tech equipment such as jet engines and rockets. We also review a more or less typical film cooling experiment reported in the literature on today's old heat transfer.

In the next few chapters, we use the methodology of the new heat transfer to redesign the above film cooling experiment and to reappraise the data. The end result is a new and better view of film cooling phenomena which little resembles the view widely accepted in the old heat transfer for almost 50 years.

THE PRACTICAL APPLICATIONS OF FILM COOLING

Hartnett, Birkebak, and Eckert (1961) discuss the practical applications of film cooling:

> A promising method for protecting a surface exposed to a high-temperature environment involves the introduction of a coolant liquid or gas through discrete slots appropriately positioned along the surface . . . This cooling scheme, commonly called film-cooling, has already found application in gas turbines and rocket nozzles and in the future may be of value in cooling leading edges of hypersonic aircraft.

Film coolant is usually injected into the boundary layer through a slot or an array of holes which attempt to direct the coolant parallel to the cooled surface. The magnitude of the cooling effect is usually described by the idea of film cooling "effectiveness" which is assigned the symbol η (no dimensions) and is defined by

$$\eta = (T_m - T_{aw})/(T_m - T_c) \qquad (25\text{-}1)$$

where T_m = temperature of mainstream fluid, F.

T_{aw} = temperature of adiabatic wall, F.

T_c = temperature of coolant at slot exit, F.

Figure (25-1) describes a film cooling scheme in which the film coolant is injected through a tangential slot.

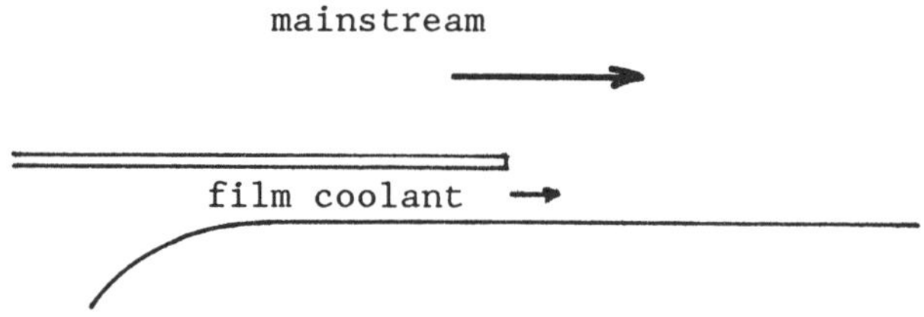

FIGURE (25-1) Tangential film cooling slot

FILM COOLING CORRELATIONS

The most widely used film cooling correlations are dimensionless power laws--the preferred correlation form in today's old heat transfer. The generic film cooling correlation form is:

$$\eta = a(x/ms)^b \qquad (25\text{-}2)$$

where x = distance downstream of slot, in.

m = ratio of W/A in slot to W/A in mainstream, no dimensions

s = slot height, in.

Kays (1966) discusses film cooling:

> . . . there is . . . a considerable body of experimental data, the results of which can be presented in a simple manner. . . . It is found that η is primarily a function of a blowing rate parameter m, the height of the injection slot s, and the distance x. . . Using air with constant free-stream velocity, Wieghardt (1946) presents the correlation
>
> $$\eta = 21.8(x/ms)^{-0.8} \qquad (25\text{-}3)$$
>
> for $0.22 < m < 0.74$ and $x/s > 100$.

A TYPICAL FILM COOLING EXPERIMENT

Hartnett, Birkebak, and Eckert (1961) performed a more or less typical film cooling experiment using the methodology of the old heat transfer. They state:

> In this study a free-stream velocity of 165 fps was used throughout the program. The injection studies were restricted to a single slot size (s = .123") and a single blowing rate (m = .28).

> . . . heated secondary air was injected from the slot and the resulting adiabatic temperature distribution was determined. The usual choice of parameters was made to represent the data. These same parameters are suggested in a semi-empirical analysis presented in the Appendix. . . . The results of the adiabatic wall temperature distribution are shown in Fig. (24), in terms of the dimensionless distance from the slot leading edge, x/ms . . . The data are correlated for x/ms > 60 by the relation
>
> $$\eta = 16.9(x/ms)^{-.8} \qquad (25\text{-}4)$$
>
> . . . the agreement of the present measurements with those of Wieghardt (1946) is fair, and both investigations are in agreement with the conclusion of the Appendix; namely, that the effectiveness is proportional to $(x/ms)^{-.8}$ at distances far downstream of the slot.

The authors plot their $\eta\{x/ms\}$ results in their Fig. 24 which is a log log graph suggested by the power law in Eq. (25-4). There is very little scatter in the points, and the authors draw a curve through them. Fig. (25-2) is similar to the authors' Fig. 24. The solid line resembles the curve drawn by the authors through the points in Fig. (24). The dashed line is Eq. (25-4).

In Fig. (25-2), notice the good agreement between experiment and correlation in the region above x/ms=100, and the disagreement in the region below this value. Because of the disagreement below x/ms=100, the authors:

- State that their deduced power law correlation is applicable only at values of x/ms greater than 60.

- Conclude that, since their deduced power law correlation and the results do not agree over the full range of the data, the data must include more than one regime.

- Conclude that their deduced power law correlation describes only the regime "far downstream of the point of injections".

The authors compare their empirical results with results obtained by other investigators using similar slots. The comparison is made in their Fig. 26 which is a log log plot of η vs x/ms. Regarding this figure, the authors state:

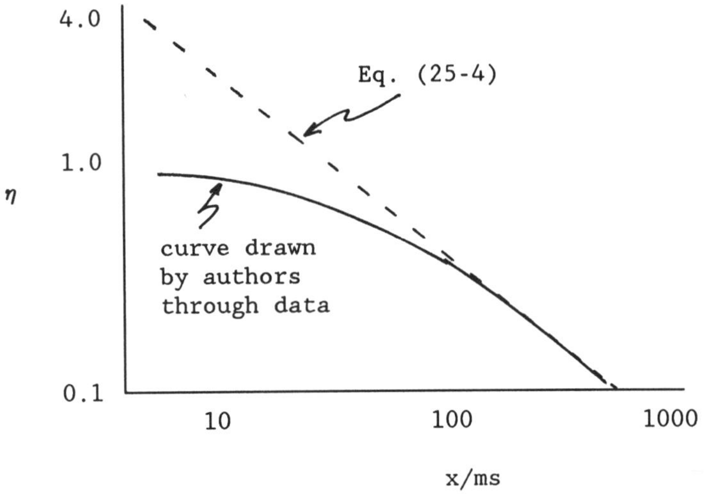

FIGURE (25-2) Sketch of Fig. 24 in Hartnett, Birkebak, and Eckert (1961)

> The effectiveness values reported by (other) investigators is shown in Fig 26 as a function of the important parameter x/ms. (This figure also shows the authors' results.) . . . A single curve can be drawn which will represent the reported values with a deviation of ±40 percent. It is clearly not possible at the present time to state the specific causes of this variation . . . (However), a reasonable estimate of the effectiveness for practical applications can be obtained from (the curve in) Fig 26.

The curve in Fig. 26 is the same curve drawn by the authors through their results plotted in Fig. 24.

CHAPTER 26

SELECTING THE IMPORTANT PARAMETERS

Selecting the parameters which determine the behavior of a specific phenomenon sounds like a simple task, but it is not. Ohm's most important contribution was his determination and quantification of the parameters volts and amps. Fourier's most important contribution was his determination that heat flux is the cardinal parameter in heat flow.

Each time we are confronted with a different phenomenon, the most important behavioral question to be answered is:

> *What are the parameters which determine/describe the phenomenon's behavior?*

This question can usually be answered correctly in several different ways, but there may be considerable difference in the quality of the answers. Some parameters may make it simple to understand/control/optimize the phenomenon, while others may make this virtually impossible.

In this chapter, we discuss the parameters of importance in film cooling. We consider the parameters selected in today's old heat transfer, and we select them using new methodology. We find that the parameters usually selected in today's old heat transfer make it virtually *impossible* to understand film cooling behavior or to optimize film cooling equipment because:

- *They result in correlations in which primary parameters are IMplicit, and secondary parameters are EXplicit!*
- *These correlations IMplicitly state that two EXplicit parameters have NO EFFECT on film cooling!!!*
- *Film cooling data generally indicate that the two explicit parameters which are implicitly described to have no effect are known to have considerable effect.*

On the other hand, it is quite simple to understand film cooling behavior and to optimize film cooling equipment using the new methodology because the important parameters appear in explicit form in the new way correlations.

SELECTING THE IMPORTANT PARAMETERS IN FILM COOLING--OLD WAY

The first step in investigating and correlating the behavior of a specific phenomenon is to identify the parameters of importance--the parameters which actually describe the behavior and will be used in the correlation.

In today's old heat transfer, there is a great deal of latitude in the selection of correlating parameters. They are usually selected by deduction, and all parameters are allowed. However, there is a preference for contrived, dimensionless parameters, and a prejudice against fundamental, dimensioned parameters in explicit form.

Correlations written in the form of dimensionless, contrived parameters are considered scientific, particularly those which are based on dimensional analysis. Correlations written in the form of dimensioned, fundamental parameters are considered unscientific, particularly if they are not dimensionally consistent.

With regard to the important parameters in film cooling, the methodology of the old heat transfer resulted in two conclusions which have been widely accepted for decades:

- "η is primarily a function of a blowing rate parameter m, the height of the slot s, and the distance x." (From Kays, (1966).)
- The first order dependence of these parameters is described by a power law in terms of the two dimensionless parameters η and x/ms. The generic form of the power law correlation is:

$$\eta = a(x/ms)^b \tag{26-1}$$

THE FLAW IN THE OLD FILM COOLING PARAMETERS

There is an important flaw in the above two conclusions about film cooling parameters:

The two conclusions are mutually exclusive!!

This flaw is very subtle. It has persisted for decades, but it must be firmly grasped by the reader before going on to the next chapter. The flaw is described in the following observations:

- If η, x, m, and s are important parameters, then Eq. (26-1) *cannot possibly describe film cooling behavior* because it implicitly states that the parameters m and s have *NO EFFECT* on η!

- Eq. (26-1) states that, if s is increased from 1 inch to 2 feet in a given application, there will be *no effect* on η, in spite of the fact that Eq. (26-1) seems to state that η would be greatly affected.

- Eq. (26-1) actually states that η depends on the parameters x, G_{ms}, and W_c/L.

    G_{ms} = mainstream W/A at the slot exit, lbs/in^{2}sec.

    W_c = coolant flow rate, pps.

    L = slot *width*, in.

It is not easy to recognize that Eq. (26-1) states that η does *NOT* depend on m and s. It is readily seen from Eq. (26-1) that, if the values of x and m are held constant, then η will vary with s. This observation is accurate, but it is not germane. The fact that Eq. (26-1) states that η depends on s at constant x and m has nothing to do with whether Eq. (26-1) states that η depends on m or s.

THE SOURCE OF THE FLAW

The source of the flaw described above is readily determined if we utilize the correlation methodology of the new heat transfer--specifically the idea that correlations should be written in terms of parameters which are both fundamental *and independent*.

If we analyze the parameters in Eq. (26-1) to determine whether or not they are independent, we find that m and s are *dependent*. This dependence is the reason Eq. (26-1) seems to state that η depends on m and s, when in fact it states that η *in no way* depends on m or s. The impact of the dependence of m on s is determined by noting:

$$m = G_c/G_{ms} \tag{26-2}$$

$$G_c = W_c/A_c = W_c/Ls \tag{26-3}$$

$$m = W_c/(G_{ms}Ls) \tag{26-4}$$

$$x/ms = xG_{ms}Ls/W_cs \tag{26-5}$$

$$x/ms = xG_{ms}L/W_c \tag{26-6}$$

Notice that the right side of Eq. (26-5) contains s in both the numerator and the denominator, and therefore the value of x/ms does NOT depend on s because s cancels. Also notice that m depends on slot area Ls, and thus when s is canceled as in Eq. (26-6), it becomes apparent that the value of x/ms does not depend on m either.

In other words, x/ms contains the parameter s explicitly in the denominator and implicitly in the numerator. If we transform the implicit s to an explicit s, it then becomes apparent that s cancels and that x/ms is independent of s. And when s is canceled, it becomes apparent that x/ms does not depend on m either!

> *It is essential to recognize that x/ms IN NO WAY depends on m or s, and that x/ms is the disguised form of* $xG_{ms}L/W_c$.

Expressions of the form $\eta\{x/ms\}$ seem to state that m and s are important, but they actually state that m and s are of *NO* importance. The confusion which results makes it virtually impossible to understand film cooling behavior, or to optimize film cooling equipment. Particularly since m has considerable importance, and s has some importance.

SELECTING THE IMPORTANT PARAMETERS IN FILM COOLING--NEW WAY

In the new heat transfer, the selection of the important parameters in film cooling is a simple matter. It is based on the following new heat transfer methodology:

- Correlations which describe Natural behavior must be written in terms of fundamental parameters (rather than contrived parameters).
- Correlation parameters must be independent.

These guidelines suggest that the important parameters in film cooling are mainstream and coolant temperatures, mainstream and coolant flow rates, and slot height and width. These parameters must be the parameters which describe film cooling behavior because they are the parameters measured in film cooling experiments intended to measure film cooling behavior.

As a practical matter, correlating parameters in the new heat transfer are usually the parameters measured in the experiment. The only parameter modification would be conversion from extensive form to intensive form--for example, from "pounds per second" to "pounds per second per square inch".

SUMMARY

In summary, the correlation methodology of the old heat transfer results in the conclusions that:

- The important parameters in film cooling are η, x, m, and s.

- Film cooling behavior is described by $\eta\{x/ms\}$.

Taken separately, each of these conclusions is possible. But taken together, *they are impossible because they are mutually exclusive.*

The methodology of the old heat transfer fails to identify the important parameters. It yields correlations in the form $\eta\{x/ms\}$ without recognizing that correlations of this form implicitly state m and s have *NO* importance! The ensuing confusion and the fact that m and s are important make it virtually impossible to understand film cooling behavior, or to design optimum film cooling equipment.

It is difficult to imagine a more undesirable correlation than one which describes behavior in terms of parameters which explicitly appear, and which are implicitly described to have no effect, even though they are known to have considerable effect. The singular cause of this undesirable result is that today's old way film cooling correlations contain the dimensionless group "m". This group contains "s" in the denominator, and therefore the product "ms" depends on neither m nor s because the implicit s in the denominator of m cancels the s which is explicit in ms.

It is essential that the reader obtain a firm grasp of this subtle and extremely important flaw. An appreciation of this flaw should leave no doubt that:

- *The correlation methodology of the old heat transfer is undesirable and must be abandoned.*

- *The old methodology must be replaced by the new methodology.*

- *Correlations should be written ONLY in terms of parameters which are both fundamental and independent.*

CHAPTER 27

A PRIORI DEDUCTION

In the old heat transfer, the purpose of a priori deduction is to determine as much as possible about parametric functionality--as much as possible about a correlation which accurately describes the behavior of the phenomenon being investigated.

This deduction always leads to correlations which are dimensionally consistent, and almost always leads to power laws written in terms of dimensionless groups. Experiment is relatively unimportant--experimental results are used primarily to establish the values of constants in deduced correlations.

In the new heat transfer, the purpose of a priori deduction does not include the deduction of parametric functionality. The literature contains ample evidence that deduction is not a reliable way to determine functionality, and therefore we do not consider deduction an acceptable way of determining functionality.

In the new heat transfer, parametric functionality is determined using Galileo's "scientific method" based on experiment and induction. The purpose of a priori deduction is simply to identify the functions which will be induced from the measured results--from the data.

Functionality with respect to a particular parameter can be determined by experiment *only* if the parameter is varied. Therefore, only those parameters which are varied over a wide range in the experiment should appear in parametric form in the resultant correlation.

For example, suppose that the important parameters are w, x, y, and z, and that we are to perform an experiment in an apparatus in which the value of z cannot be varied over a wide range. Since z cannot be varied, the experiment cannot tell us anything about functionality with respect to z, and therefore z should not appear in parametric form in the correlation induced from the experiment. It should appear only in numeric form, if at all.

(It may appear in parametric form in a later revision of the correlation which is generalized on the basis of deduction or similarity to other experiments, but that is a separate subject which need not concern us here. Here we are concerned only with the correlation which is the direct result of the experiment.)

Knowing that w and x will be varied in the experiment, we deduce that the desired correlation is described by

$$y = f\{x\}\ f\{w\} \qquad (27\text{-}1)$$

where the f's refer to unknown functions. Experimental results are used to quantitatively induce the unknown functions in Eq. (27-1), and that is why experiment design is an important part of the new methodology.

In this chapter, we discuss a priori deduction in the old and new heat transfers. The discussion is specific in that we consider the a priori deduction for a film cooling experiment, but we are little concerned with film cooling per se. Our real interest is in the generic methodology of a priori deduction. Film cooling is merely the vehicle used for illustration.

The film cooling publication by Hartnett, Birkebak, and Eckert (1961) is used to exemplify a priori deduction in the old heat transfer. Then we perform a priori deduction for the same film cooling experiment using the new heat transfer.

The merit of a priori deduction depends in part on how well it agrees with experimental results. Therefore we must postpone judgement about the relative merit of old/new methodology until we have reviewed the experimental results in a later chapter. Following that review, we will return to the subject of a priori deduction. We will find that the correlation based on the old way deduction bears little resemblance to the data it purports to describe. We will logically conclude that the old methodology is unreliable. And that the new methodology is better than the old.

A PRIORI DEDUCTION FOR A FILM COOLING EXPERIMENT--OLD WAY

The typical role of a priori deduction in the old heat transfer is well illustrated in the experiment by Hartnett, Birkebak, and Eckert (1961). Their deduction is described in the following excerpts:

> The usual choice of parameters (η and x/ms) was made to represent the data. These same parameters are suggested in a semi-empirical analysis presented in the Appendix.

> . . . the agreement of the present measurements with those of Wieghardt is fair, and both investigations are in agreement with the conclusion of the Appendix: namely, that the effectiveness is proportional to $(x/ms)^{-.8}$ at distances far downstream from the slot.

In the Appendix, the authors deduce that film cooling behavior is described by an η correlation in which x/ms appears with an exponent of −.8. They point out that this result "agrees with the theoretical results of Klein and Tribus (1953)", and they conclude that the data will correlate in the form

$$\eta = a(x/ms)^{-.8} \qquad (27\text{-}2)$$

The authors deduce the film cooling correlation in its entirety with the single exception of the value of "a". We will not review the authors' deduction. Our interest lies in the fact that Eq. (27-2) was deduced *before* the experiment. We have no interest in *how* it was deduced.

A PRIORI DEDUCTION FOR A FILM COOLING EXPERIMENT--NEW WAY

The purpose of a priori deduction in the new heat transfer is to identify the unknown functions which will be induced from the data. The parameters which are to be varied in the experiment are listed, and a generic correlation is deduced which includes all the parameters on the list. The correlation is written in terms of unknown functions and is based on assumed separability. The experiment is designed to provide the data required to establish the unknown functions, and to verify the assumption of separability.

A film cooling experiment in which the flow path is rectangular and the coolant is injected through a tangential slot would include at least the following data:

Flow path height and width.

Slot height and width.

Coolant flow rate and temperature.

Mainstream flow rate and temperature.

Wall temperature distribution downstream of the slot, $T_w\{x\}$, where w refers to an insulated wall assumed to be adiabatic. x is the distance downstream of the slot, inches.

A priori deduction in preparation for a film cooling experiment to be performed at a mainstream temperature of 650 F would typically result in an expression similar to:

$$T_w\{x\} = 650 - f\{x\}\ f\{W_c'\}\ f\{G_m\}\ f\{s\}\ f\{650 - T_c\} \qquad (27\text{-}3)$$

Notice that we have made NO deduction about functionality. We have merely identified the unknown functions which will be induced from the experimental results. We refuse to deduce these functions because deduction is not a reliable way to determine functionality. We consider that the determination of functionality is the true purpose of experiment, and we intend to design the experiment so that it will in fact provide the data required to establish the unknown functions in Eq. (27-3).

It is important to note that Eq. (27-3) is based on the assumption of separability--the assumption that each of the parameters acts independently--ie the assumption that the effect of each parameter in no way depends on the level of the other parameters. This assumption is an integral part of the old heat transfer, but it is usually made implicitly rather than explicitly.

This assumption greatly simplifies the task of determining the correlation because it allows us to substitute the specific form of Eq. (27-3) for the more general form of Eq. (27-4):

$$T_w\{x\} = 650 - f\{x, W_c', G_m, s, (650 - T_c)\} \qquad (27\text{-}4)$$

Separability is indicated in most correlations of the old heat transfer, including the film cooling correlations by Wieghardt, Eq. (25-3), and by Hartnett, Birkebak, and Eckert, Eq. (25-4). Separability is readily demonstrated by experiment, and we intend to do so.

If the experiment demonstrates that separability does not apply to the phenomenon of interest, the deduced equation must be revised to indicate that separability does not apply. The nature of the revision can often be determined by examining the data. If not, it may be necessary to redesign the experiment as described in the first edition of this book.

In summary, a priori deduction in the new heat transfer would typically result in an expression resembling Eq. (27-3). This equation identifies the unknown functions to be determined by experiment and induction. In the old heat transfer, these unknown functions would be determined by a priori deduction such as the deduction which leads to Eq. (27-2). In a very real sense, a priori deduction in the old heat transfer is replaced by experiment and induction in the new heat transfer.

OLD WAY VS NEW WAY

The old heat transfer methodology results in "knowing" a great deal about the experimental correlation *before the experiment even begins*. In the film cooling example, Eq. (27-2) was completely determined a priori with the one exception that a constant in the correlation could not be deduced. The experiment was needed merely to obtain the data required to establish the value of that one constant.

If Eq. (27-2) accurately describes film cooling behavior, then this example demonstrates that the old way is better than the new way. It is much simpler/cheaper/quicker to determine functionality by deduction rather than by experiment and induction.

On the other hand, if Eq. (27-2) does not accurately describe film cooling behavior, then the old methodology is obviously undesirable and should be abandoned. The danger is that Eq. (27-2) might be used to design equipment, and a poor and possibly disastrous design would result if Eq. (27-2) did not accurately describe film cooling behavior.

In order to readily compare Eqs. (27-2) and (27-3), we must first rewrite Eq. (27-2) in terms of the fundamental parameters which were replaced by the dimensionless groups η and x/ms. The resultant expression is

$$T_w = T_m - a(x)^{-.8}(W_c')^{.8}(G_m)^{-.8}(T_m - T_c) \qquad (27\text{-}5)$$

Eq. (27-5) is *identical* to (27-2). The one difference is in form--there is no difference in substance. Eq. (27-5) is in the form of fundamental and independent parameters--the clear and simple language of the new heat transfer. Eq. (27-2) is in the form of dimensionless groups--the opaque and confusing expertise of the old heat transfer.

Eq. (27-5) probably looks strange to anyone who has used dimensionless groups since her/his undergraduate days. But notice that it is transparent--it readily reveals that m and s have nothing to do with film cooling. Also notice that Eq. (27-2) is so opaque that, for 50 years, it has generally concealed the fact that *it actually states m and s have nothing to do with film cooling!*

Compare Eqs. (27-5) and (27-3) and notice that Eq. (27-3) makes no statement about the unknown functions. It simply notes that these functions must somehow be determined in order to describe the behavior of film cooling. Also note that Eq. (27-5) contains a great deal of information about the unknown functions in Eq. (27-3). In particular, Eq. (27-5) states that:

- f{x} is a power law with exponent −0.8.
- f{W_c'} is a power law with exponent 0.8.
- f{G_m} is a power law with exponent −0.8.
- f{s} is a constant--ie s has no effect on T_w.
- f{m} is a constant--ie m has no effect on T_w.
- f{$T_m - T_c$} is a power law with exponent 1.0.

In summary, the old methodology approaches experiment with faith--faith in a priori deduction, faith in power laws, and faith that the data will resemble the correlation deduced a priori. The new methodology approaches experiment with ignorance and confidence--ignorant of parametric functionality, and confident that the experiment can be designed to determine functionality.

As the preceding chapter and the next several chapters demonstrate, the old heat transfer faith in a priori deduction is poorly founded. The deduced correlation, Eq. (27-2), bears *no resemblance* to the real world behavior of film cooling!

The old heat transfer faith in a priori deduction is better replaced by the a priori ignorance and confidence in experiment found in the scientific method of Galileo--the base line method of the new heat transfer.

CHAPTER 28

EXPERIMENT DESIGN AND SEPARABILITY

In today's old heat transfer, the purpose of experiment has little to do with the determination of functionality. Functionality is largely determined by a priori deduction, usually by the deduction called "dimensional analysis".

The deduced correlations are generally power laws written in terms of dimensionless groups. The dimensionless groups replace/represent/substitute for the fundamental parameters in the experiment, and consequently experiment design is concerned with dimensionless groups rather than fundamental parameters.

The purpose of experiment in the old heat transfer is to provide data which is used only to establish the values of pure constants in deduced correlations. This rather secondary purpose requires only a small amount of data, and the design of such experiments is a trivial matter which receives little conscious attention.

In the new heat transfer, the purpose of experiment is to provide data which is used to establish functionality among the fundamental parameters which dictate the behavior of the phenomenon being investigated. This requires a large amount of data, and the design of such experiments is an important matter which receives a great deal of attention.

In the new heat transfer, experiment design is usually based on the simplifying assumption of *separability*. In other words, we assume that the fundamental parameters act *independently* on the phenomenon being investigated. For example, given that y depends on x and z, we assume that the specific form of Eq. (28-2) may be substituted for the general form of Eq. (28-1):

$$y = f\{x,z\} \qquad (28\text{-}1)$$

$$y = f\{x\}\, f\{z\} \qquad (28\text{-}2)$$

The assumption of separability is implicit in much of the old heat transfer methodology, and therefore we anticipate that it will often be valid. Since separability is readily verified by experiment, experiments are usually designed to provide the data which tests for separability.

The separability assumption greatly simplifies both the design and performance of experiments by allowing us to *individually* determine the functionality of each parameter in the experiment. This is accomplished in a set of small, simple experiments designed so that the parameter of interest is varied while the other parameters are maintained at fixed values.

> *The end result is that separability allows us to substitute a set of simple, one-dimensional experiments for the complex, multi-dimensional experiment which actually confronts us.*

Since separability is a questionable assumption, it must be verified. This is accomplished by demonstrating that the measured functionality of each parameter is independent of the level of the other parameters.

In this chapter, we compare experiment design in the old and the new heat transfers. We again use film cooling as the vehicle for illustration, and we again use the publication by Hartnett, Birkebak, and Eckert to exemplify today's old heat transfer methodology.

THE DESIGN OF A FILM COOLING EXPERIMENT--OLD WAY

The design of the film cooling experiment by Hartnett, Birkebak, and Eckert (1961) is described in the following excerpts:

> In this study a free stream velocity of 165 fps was used throughout the program.
>
> The injection studies were restricted to a single slot size (.123") and a single blowing rate (m =.28).
>
> The temperature of the injected air . . . was varied from 10 to 150 deg F above the temperature of the free stream (which was approximately 70 F).

It is important to note that this experiment is designed to obtain data at:

- *One value* of mainstream flow rate (165 fps).
- *One value* of mainstream temperature (70 F).
- *One value* of coolant flow rate. (The flow rate is not given explicitly. It is given implicitly in the value of the dimensionless group called "blowing rate parameter" and assigned the symbol "m".)
- *One value* of the dimensionless group m (m =.28).
- *One value* of slot height (s =.123").
- A range of x values from 0 to 17".
- A range of coolant temperatures from 80 to 220 F. (Notice that the "coolant" is hotter than the mainstream, and thus the experiment is actually a film heating experiment. Presumably the a priori deduction indicated that film heating and film cooling are equal and opposite.)

Because x is varied over a wide range, this experiment design includes the desired variation in "the important parameter x/ms". But also notice that this design involves *no variation* in most of the fundamental parameters which are important in film cooling--no variation in mainstream flow rate, no variation in coolant flow rate, no variation in mainstream temperature, no variation in slot size.

Because we deduce that the influence of the fundamental parameters is exerted through "the important parameter x/ms", we conclude that this group is the "parameter" which must be varied in the experiment. The manner in which this group is varied is of no importance, and therefore the group is varied in whatever manner is most convenient.

For example, it is not necessary to individually vary x, m, and s. It is sufficient for our purposes to vary only x--to perform the entire experiment at only one value of m and one value of s. This is the most convenient way to vary the value of x/ms, and this is in fact the way the experiment by Hartnett, Birkebak, and Eckert was designed and performed.

GENERIC EXPERIMENT DESIGN--NEW WAY

In the new heat transfer, experiment design is based on the following line of reasoning:

- Deduction can be used to predict the functional relationships among Natural parameters. But deduction is not reliable. It often results in predicted behavior which does not resemble actual behavior. And we require correlations which accurately describe actual behavior.

- There is only one reliable way to obtain correlations which accurately describe actual behavior. That one way is experiment and induction--the "scientific method" of Galileo.

- If we wish to experimentally determine the functionality of a particular parameter, that parameter must be varied over a wide range in an experiment.

- In order to vary parameters over the widest possible range, an experiment design should explore the complete operating envelope of the experimental facility.

- The task of designing experiments is greatly simplified if it is assumed that parametric effects are separable. Since this assumption is implicit in much of the old heat transfer, it seems likely to be valid in the new heat transfer. Therefore it is reasonable to design experiments based on the assumption of separability *provided* the design includes verification of the assumption.

The assumption of separability allows us to individually determine the functionality of each parameter from a small and simple experiment in which one parameter is varied while the others are maintained at fixed values. Separability is readily verified by demonstrating that the measured functionality of each parameter is independent of the level of the other parameters.

> *The end result is that separability allows us to substitute a set of simple, one-dimensional experiments for the multi-dimensional experiment which actually confronts us.*

The assumption of separability is open to question, but it is implicit in so much of the old heat transfer that it seems likely to apply in much of the new heat transfer. If the data indicates that separability does not apply, the equation deduced a priori must be revised. It may also be necessary to redesign the experiment.

THE DESIGN OF A FILM COOLING EXPERIMENT--NEW WAY

Let us design a film cooling experiment for a facility described by the following (100% indicates the limit of the facility's operating envelope):

Coolant flow rate: Variable, 0 to 100%.

Mainstream flow rate: Variable, 0 to 100%.

Mainstream temperature: Fixed, 450 F.

Coolant temperature: Variable, 100 to 450 F.

Distance from injection point: Variable, 0 to 100%.

Injection geometry: Tangential slot.

Slot height: Variable, 0 to 100%.

If we assume separability, we can determine each of the unknown functions in Eq. (27-3) by varying that parameter while the others are held constant. The experiment then becomes a set of small, simple experiments, each of which is designed to generate the data necessary to induce one of the unknown functions in Eq. (27-3).

$$T_w\{x\} = 450 - f\{x\}\ f\{W_c'\}\ f\{G_m\}\ f\{s\}\ f\{450-T_c\} \qquad (27\text{-}3)$$

For example, we can determine f{x} from a single test point in which the other parameters are maintained at 50%. In order to verify the assumption of separability, we should determine f{x} with the variable parameters maintained at more than one level. Therefore, let us design the f{x} experiment to include test points in which the other parameters are maintained at 25, 50, 75, and 100%.

At each test point, f{x} is determined by noting that

$$T_w\{x\} = 450 - a_i f\{x\} \qquad (27\text{-}4)$$

$$f\{x\} = (450 - T_w\{x\})/a_i \qquad (27\text{-}5)$$

where a_i has a different value in each of the four test points. If there is no significant difference in f{x} induced from each test point, we will have demonstrated that separability applies to the x effect on film cooling.

If there is a significant difference in the f{x} result from the four test points, separability does not apply to the x effect. In this event, the deduced equation must be revised to indicate the lack of separability. If the required revision can not be determined by inspection of the data, it will be necessary to redesign the experiment in order to address the interaction between x and the other parameters--ie to determine how the x effect depends on the level of each of the other parameters.

Similarly, we can determine $f(W_c')$ by varying coolant flow rate from 0 to 100% in steps of 10% while the other parameters are maintained at 25%, then repeating the series of test points with the other parameters maintained at 50, 75, and 100%. The separability analysis is the same as described above with regard to f(x).

SUMMARY

The particular advantage of assuming separability is that it greatly simplifies the experiment and the data reduction by allowing us to transform a complex experiment in many variables into an array of one variable experiments, each of which is simple to perform and simple to correlate.

It is important to note that separability is not assured, and therefore it must be demonstrated whenever it is used to simplify an experiment. However, since most correlations of the old heat transfer indicate that the variables act independently, it is reasonable to expect that separability will usually apply.

If the experiment results indicate that separability does not apply, the equation deduced a priori must be revised, and it may also be necessary to redesign the experiment. The redesign and the resultant data analysis are described in the first edition of this book.

CHAPTER 29

DATA REDUCTION AND CORRELATION

In today's old heat transfer, data reduction involves the conversion of raw data to dimensionless groups determined by a priori deduction. Correlation involves plotting the dimensionless groups on log log paper, drawing a straight line, and using the line to determine the constants in an assumed power law.

The end result of this methodology is that data correlation is largely replaced by a priori deduction and the assumption of power law behavior. The data are used merely to establish the value of constants in the assumed power law. *By no stretch of the imagination can this be regarded as data correlation.*

In the new heat transfer, raw data are reduced to fundamental parameters--to flow rates and temperatures and heat fluxes, and pressures. These fundamental, dimensioned parameters retain their identity *even in the resultant correlations*.

Correlation involves plotting the measured behavior of fundamental parameters on linear paper in order to induce parametric correlations which contain only fundamental parameters.

In this chapter, we compare data reduction and correlation in the old and new heat transfers. As before, we use film cooling to illustrate the old/new methodology, and we rely on the publication by Hartnett, Birkebak, and Eckert to describe the methodology of the old heat transfer.

DATA REDUCTION AND CORRELATION--OLD WAY

The film cooling experiment by Hartnett, Birkebak, and Eckert is typical of data reduction and correlation in the old heat transfer. It includes the following elements:

- Reduce the data to dimensionless groups determined by *a priori* deduction.
- Plot the dimensionless groups on log log paper.
- Draw the best straight line through all the points or through some region of the points--some invented regime.
- Use the straight line to determine the value of the constants in an assumed power law.

The above elements are apparent in the following steps described by Hartnett, Birkebak, and Eckert:

- The data were reduced to the dimensionless groups η and x/ms.
- The dimensionless group results were plotted on log log paper in the authors' Fig. 24. (Fig. (25-2) resembles Fig. 24).
- A curve was drawn through the points.
- A region of the curve was approximated by a straight line segment.
- The equation of the straight line segment was used to determine the value of "a" in Eq. (29-1) which had been obtained by a priori deduction:

$$\eta = a(x/ms)^{-.8} \qquad (29\text{-}1)$$

From the straight line segment, the authors concluded that Eq. (29-2) describes film cooling from a tangential slot in the region far from the point of injection:

$$\eta = 16.9(x/ms)^{-.8} \qquad (29\text{-}2)$$

USING THE NEW HEAT TRANSFER TO CORRELATE THE RESULTS BY HARTNETT, BIRKEBAK, AND ECKERT

Now let us use the methodology of the new heat transfer to correlate the results by Hartnett, Birkebak, and Eckert:

- Since only the parameters x and "coolant" temperature were varied in the experiment, these are the *only* parameters which should appear in parametric form in the experimental correlation.

- Deduction before the experiment resulted in Eq. (29-3) in which "70" is the numeric value of T_m.

$$T_w = 70 + f\{x\}\ f\{T_c - 70\} \qquad (29\text{-}3)$$

- Eq. (29-3) is based on the assumption of separability. Since separability is indicated in the authors' result, Eq. (29-1), this assumption has no impact on the relative merit of old/new methodology.

- There is little scatter in the authors' Figure 24 which describes $\eta\{x/ms\}$ at a mainstream temperature of 70 F and "coolant" temperature levels of 80 to 220 F. The lack of scatter demonstrates that η is independent of coolant temperature in the range tested. Therefore the film heating effect is proportional to the ΔT between coolant and mainstream. (The lack of scatter also demonstrates that the effects of coolant temperature and x are separable.) Therefore we may conclude from the experiment that $f\{T_c-70\}$ in Eq. (29–3) is given by Eq. (29-4) in which "a" is a pure constant:

$$f\{T_c - 70\} = a(T_c - 70) \qquad (29\text{-}4)$$

- The authors' Eq. (29-1) indicates that f{x} is given by Eq. (29-5) in which "b" is a pure constant:

$$f\{x\} = bx^{-.8} \qquad (29\text{-}5)$$

- The fact that $\eta\{x/ms\}$ is not a straight line in the authors' Fig. 24 indicates that the data do *NOT* support the power law described by Eq. (29-5). Moreover, Eqs.

(29-1) and (29-5) state that η equals infinity at x equal to zero, whereas it is not possible for η to be greater than unity! (Since it is not possible for the cooled wall to be cooler than the coolant--or rather for the heated wall to be hotter than the heating fluid.)

- Since the data do not support the power law in Eq. (29-5), we conclude that there is no good reason to plot the results on log log paper. We replot the points shown on the authors' Fig. 24, this time on linear paper. From this new way plot, we induce the function in Eq. (29-6).

$$f\{x\} = 1/(1 + .412x) \qquad (29\text{-}6)$$

As the reader may readily verify by plotting Eq. (29-6) on Fig. 24 in the subject article, Eq. (29-6) essentially coincides with the curve drawn by the authors through the points in Fig. 24. In other words, this function closely agrees with *all the points* in Fig. 24, not merely the points above x/ms=100 (the region where the authors' correlation agrees with their experimental results).

(Since the authors' Fig. 24 is f{x/ms} rather than f{x}, Eq. (29-6) must be transformed to f{x/ms} by noting that the experiment was performed with an m value of 0.28 and an s value of 0.123. In the transformed expression, .0142x/ms replaces .412x.)

- Substituting Eqs. (29-4) and (29-6) into (29-3), we obtain the experimental correlation:

$$T_w = 70 + (T_c - 70)/(1 + .412x) \qquad (29\text{-}7)$$

This dimensional, non-power-law correlation is qualified by the restriction that it is reliable only if x is in the range 0 to 17 inches, T_c is in the range 80 to 220 F, the slot height is .123", the mainstream temperature is 70 F, the mainstream velocity is 165 fps, the mainstream pressure is ambient, the coolant flow rate corresponds to an "m" value equal to 0.28, and the slot geometry of interest closely resembles the slot geometry in the experiment.

- Eq. (29-7) is a new way experimental correlation. It is accurate and reliable because it closely agrees with the data, but it is not very useful because it describes only the effect of x and coolant temperature on film heating. It is important to note that this severe drawback is the direct result of the limited scope of the subject experiment. It has nothing to do with old/new methodology.

If we are willing to accept without reliable proof that η and x/ms are the important parameters, and also that film heating and film cooling are equal and opposite, we may generalize Eq. (29-7) and obtain Eqs. (29-8) and (29-9):

$$\eta = 1/(1 + .0142x/ms) \qquad (29\text{-}8)$$

$$T_w = T_m - (T_m - T_c)/(1 + .0142xG_m/W_c') \qquad (29\text{-}9)$$

Notice that Eqs. (29-8) and (29-9) are identical. Eq. (29-8) is written in the opaque language of the old heat transfer; Eq. (29-9) is written in the transparent language of the new heat transfer.

(The relative opacity/transparency of these equations is demonstrated by using them to determine the optimum value of slot height (s) for a given application--ie the value of s which will give the coolest wall temperature for a given coolant flow rate. Inspection of Eq. (29-9) indicates that s has nothing to do with wall temperature, and therefore all values of s are "optimum". The reader is invited to obtain this same correct result from Eq. (29-8).)

It is important to recognize that Eq. (29-7) represents what we know about film cooling, whereas Eqs. (29-8) and (29-9) represent extensions of what we know based on what we think--ie based on deduction. And since deduction is not reliable, we must consider that Eqs. (29-8) and (29-9) are not reliable.

In summary:

- If the methodology of the new heat transfer is used to correlate the results of the experiment by Hartnett,

Birkebak, and Eckert, Eq. (29-7) is the experimental and reliable conclusion.

- If we are willing to accept without reliable proof that η and x/ms are the important parameters, and that film cooling and film heating are equal and opposite, we can generalize this result and obtain Eq. (29-8) (old way) or (29-9) (new way).

- Eqs. (29-8) and (29-9) are not reliable because they are largely based on deduction. If they are used for equipment design and analysis, a poor design is a more than likely result.

DATA REDUCTION AND CORRELATION--NEW WAY

Now let us forget the film heating experiment by Hartnett, Birkebak, and Eckert, and consider the more comprehensive film cooling experiment designed in the preceding chapter.

Recall that a priori deduction resulted in Eq. (27-3):

$$T_w\{x\} = 450 - f\{x\}\ f\{W_c'\}\ f\{G_m\}\ f\{s\}\ f\{450-T_c\} \qquad (27\text{-}3)$$

The experiment design calls for a small experiment for each unknown function in Eq. (27-3). Since Eq. (27-3) is based on the questionable assumption of separability, each small experiment must test this assumption.

The data reduction task is to reduce the raw data to fundamental parameters such as flow rates, temperatures, heat fluxes, volts, etc.

The correlation task is to induce each of the functions in Eq. (27-3) from the small experiments designed for that purpose. Since we have no data, let us discuss this in a general way, beginning with f{x}.

Recall that the small experiment to determine f{x} involved test points at 25, 50, 75, and 100%. Eq. (27-3) indicates that, for these test points, the relationship between wall temperature and x is given by Eq. (29-10) where a_i is a pure constant within each test point, and the i subscript refers to test point:

$$T_w\{x\} = 450 - a_i f\{x\} \qquad (29\text{-}10)$$

Substituting the measured value of mainstream temperature for "450", and normalizing f{x} to its value at x=0, we obtain Eq. (29-11) at each of the four test points:

$$f\{x\} = (T_m - T_w\{x\})/(T_m - T_w\{x=0\}) \qquad (29\text{-}11)$$

Eq. (29-11) will give the same result for each of the four test points in the f{x} experiment if the effect of the x parameter is truly separable.

Therefore, if we plot the f{x} results from each of the four test points on the same graph, and if little scatter results, the lack of scatter will demonstrate that the effect of the x parameter is truly separable from the effects of the other parameters.

If the f{x} results do not resemble each other, the lack of resemblance demonstrates that the x parameter is *NOT* separable and requires that Eq. (27-3) be modified to agree with the results--ie to indicate that the x effect is not separable.

(Because of the limited scope of the experiment by Hartnett, Birkebak, and Eckert, there is insufficient data to determine whether or not the x effect is separable as suggested by their correlation, Eq. (29-2). More comprehensive experiments reported in the literature indicate that the x effect is *NOT* separable.)

Assuming the f{x} functions resemble each other, we inspect the graph in order to induce an analytic expression for f{x}. The inductive process is aided by considering the values of f{x} at x values of zero and infinity, and requiring that the induced function agree with these endpoint values.

To illustrate, Eq. (29-11) states that $f\{x\}=1$ at $x=0$. Since the film cooling effect can extend for only a finite distance downstream of the injection point, it is obvious that $f\{x\}=0$ at $x=\infty$. In order to obtain agreement with these endpoints, we require an analytic expression which gives

$$f\{x = 0\} = 1 \tag{29-12}$$

$$f\{x = \infty\} = 0 \tag{29-13}$$

Eq. (29-14) is a simple analytic expression which agrees with both Eqs. (29-12) and (29-13):

$$f\{x\} = 1/(1 + ax) \tag{29-14}$$

As shown in the preceding section, Eq. (29-14) closely agrees with the $f\{x\}$ results reported by Hartnett, Birkebak, and Eckert. It is included in the new way correlation of their results, Eqs. (29-7) and (29-9).

The induction of the unknown function $f\{W_c'\}$ is somewhat more involved than the induction of $f\{x\}$ because several test points are required to describe $f\{W_c'\}$. Recall that the experiment design calls for stepwise increases in coolant flow rate while the other parameters are maintained at 25, 50, 75, and 100%. For these test points, Eq. (27-3) gives Eq. (29-15) where the w subscript refers to wall, x refers to the value of x, and i refers to the level of the other parameters:

$$T_{w,x,i}\{W_c'\} = 450 - b_{x,i}f\{W_c'\} \tag{29-15}$$

Substituting the measured value of T_m for 450 and normalizing $f\{W_c'\}$ to its value at $W_c'=100\%$, we obtain

$$f\{W_c'\}_{x,i} = (T_m - T_{w,x,i}\{W_c'\})/(T_m - T_{w,x,i}\{W_c'=100\%\}) \tag{29-16}$$

As before, we plot the several $f\{W_c'\}_{x,i}$ results on one linear graph. Assuming the functions resemble each other, we draw a curve through the points and induce an analytic expression for the curve. Since there can be no cooling unless there is cooling flow, the expression must indicate that $f\{W_c' = 0\} = 0$.

(We cannot use the experiment by Hartnett, Birkebak, and Eckert to determine $f\{W_c'\}$ because coolant flow rate was not varied in that experiment.)

We continue in this manner until all unknown functions in Eq. (27-3) have been induced. Substituting the induced functions into Eq. (27-3) results in an experimental and reliable correlation. This correlation truly describes what was learned from the experiment. The correlation may be extended by deduction, but it must be borne in mind that deduction is unreliable, and therefore correlations are unreliable insofar as they are based on deduction.

CONCLUSIONS

The data reduction methodology of the old heat transfer is undesirable because it converts data to dimensionless groups, and this conversion effectively destroys the information in data.

The data reduction methodology of the new heat transfer is desirable because it recognizes the cardinal importance of data--because the conversion of fundamental parameters to group parameters is not permitted--because group parameters are *abandoned/forbidden/outlawed*.

The data correlation methodology of the old heat transfer is undesirable because it substitutes a priori deduction and power law assumption for data correlation.

The data correlation methodology of the new heat transfer is desirable because it is in fact data correlation and not deduction and assumption. *It is the scientific method of Galileo.*

CHAPTER 30

EXPERIMENTAL CORRELATIONS AND SCATTER

In today's old heat transfer, correlations have much to do with preconceived ideas and a priori deduction, and little to do with experimental results--little to do with data.

In the new heat transfer, we recognize two types of correlations: *experimental correlations* which are the result of experiment and induction; *design correlations* which are the combined result of experiment and deduction.

In this chapter, we consider experimental correlations and design correlations. We review the correlation conclusions reached from a specific experiment using old heat transfer methodology. Then we use new heat transfer methodology to reach correlation conclusions *from the same experiment.* The conclusions are quite different.

We also address the subject of scatter. We find that much of the large scatter about the old way correlation was due to confusion caused by one of the group parameters in the correlation.

We again use a film cooling experiment as the vehicle for illustration, and we again rely on the publication by Hartnett, Birkebak, and Eckert (1961) to describe today's old heat transfer methodology.

We find that the methodology of the old heat transfer suffers from several faults not shared by the new methodology. In particular, we find that the old methodology can and does result in:

- Correlations which do not resemble the data they purport to describe.

- Correlations which seem to account for parameters they do *NOT* account for.
- Unnecessary scatter which is the direct result of confusion caused by a group parameter in the correlation.

CORRELATIONS

In the old heat transfer, no distinction is generally made between correlations which are largely the result of experiment, and correlations which are the largely the result of deduction.

When an experiment has been performed in the old heat transfer, the first task is to reduce the data to the "important dimensionless parameters". No data analysis or correlation is performed until the data have been replaced/represented by dimensionless parameters.

Because of this substitution, we necessarily lose sight of the fundamental parameters in the experiment, and we correlate in a way which is completely general, even if the experiment was of limited scope. The end result is that no distinction is made between that part of the correlation which is based on experiment, and that part which is based on deduction.

In the new heat transfer, the first task is to reduce the data to fundamental parameters such as flows and pressures and temperatures and heat fluxes. These parameters are analyzed to obtain "experimental correlations":

> *Experimental correlations* are induced from data. They contain *ONLY* fundamental parameters varied over a *wide, specified range* in the experiment.

Since experiments are usually of limited scope, experimental correlations lack generality and are seldom suitable for design and analysis. In order to obtain acceptable design correlations, it is often necessary to generalize experimental correlations by deduction.

> *Design correlations* are based on a combination of experiment and deduction. They contain only fundamental parameters, and they are presented with a specification of the range in which they are reliable--ie the range of the parameters in underlying experiment(s).

CORRELATION CONCLUSIONS--OLD WAY

Hartnett, Birkebak, and Eckert (1961) performed a film cooling experiment from which they obtained a generalized correlation written in dimensionless parameters. The authors state:

> The data are correlated for $x/ms \geq 60$ by the relation
>
> $$\eta = 16.9(x/ms)^{-.8} \tag{30-1}$$

Notice that this correlation is presented for general use with a single restriction--it is to be used only in the regime far from the slot--ie only where "$x/ms \geq 60$".

The authors note that

> The theoretical analysis by Klein and Tribus (1953) (results in a correlation which is identical to Eq. (30-1) except that the analysis) gives a constant of 21.2 for a slot Reynolds number of 2100.

Since the correlation based on experiment closely agrees with a correlation based on theory, there is good reason to conclude that Eq.(30-1) describes film cooling behavior in a general way with reasonable accuracy in the regime far from the slot.

CORRELATION CONCLUSIONS--NEW WAY

In the last several chapters, we used the new methodology to analyze the experimental results reported by Hartnett, Birkebak, and Eckert (1961). This analysis supports the following correlation conclusions:

- The data are correlated for x/ms ≥ 0 by the relation

$$\eta = 1/(1 + .0142x/ms) \qquad (29\text{-}6)$$

- Eq.(29-6) is not an experimental correlation because it is based primarily on deduction. Eq.(29-6) is a design correlation.

- There is no experimental basis for the m functionality in Eq.(29-6) because m was not varied in the experiment. The indicated m functionality is solely the result of deduction.

- There is no experimental basis for the s functionality in Eq.(29-6) because s was not varied in the experiment. The indicated s functionality is solely the result of deduction.

- The x functionality in Eq.(29-6) was induced from the data and in fact agrees with the data *over the entire range of the data*. Therefore Eq.(29-6) indicates that there is only one "regime" in the subject experiment.

  Recall that the x functionality in the authors' Eq.(30-1) was deduced a priori and in fact agrees with the data only in the regime "far from slot". In the regime "near to slot", there is no agreement. For example, at (x=0), the data indicate η equals *unity*, whereas Eq.(30-1) indicates η equals *infinity*!

  In essence, the "near slot" and "far from slot" regimes were invented in order to excuse the fact that the power law correlation in Eq.(30-1) does not agree with the x functionality indicated by the data. Since the new methodology results in Eq.(29-6) which *does* agree with the x functionality described by the data, there is no need to invent regimes in the new heat transfer view of these experimental results.

- The form of Eqs.(29-6) and (30-1) is highly undesirable. Both equations seem to state that m and s have strong influence on η, but they actually state that m and s have *NO* influence on η. The cause of this seeming anomaly is that m is inversely proportional to s, and therefore the product ms is independent of m and s.

  This difficulty is removed by replacing the old way terminology in Eq.(29-6) with new way terminology--by replacing the dimensionless groups η and m with fundamental, independent parameters such as temperature and flow rate. The result is Eq. (29-7), a new way design correlation based on the subject experiment and the deduction which indicates that x/ms is "the important dimensionless group":

$$T_w = T_m - (T_m - T_c)/(1 + .0142xG_m/W_c') \qquad (29\text{-}7)$$

  Eq.(29-7) is identical to Eq. (29-6) except for form. Notice that the form of Eq.(29-7) is better because it readily reveals what the designer/analyst wants to know:

  How are the fundamental parameters related to each other?

  For example, Eq.(29-7) readily reveals that the film cooling effect is independent of m and s, whereas Eq. (29-6) seems to suggest that m and s have first order importance.

- Eq.(29-6) states, in a very opaque way, that the film cooling effect is independent of m. However, other experiments in which m was varied generally indicate that the film cooling effect *does* depend on m.

  For example, in the discussion following the publication by Hartnett, Birkebak, and Eckert (1961), Seban points out that the experimental results in Seban (1960) included variation in m and were correlated in the form

$$\eta = 25\ m^{.4}(x/ms)^{-.8} \qquad (30\text{-}2)$$

  Since x/ms in no way depends on m or s, Eq.(30-2) indicates that η is proportional to $m^{.4}$.

- Eq.(29-6) states, in a translucent way, that the film *cooling* effect is proportional to $(T_m - T_c)$. This conclusion is supported by film *heating* data in which the temperature difference was varied from 10 to 150 F.

  But practical film *cooling* temperature differences are much larger--often of the order of 1000 F or more. And experiments conducted at large temperature differences generally indicate that the film cooling effect is *NOT* proportional to practical temperature differences.

  For example, Fig. 26 in the subject publication compares the film cooling results obtained by Hartnett, Birkebak, and Eckert with results obtained by several other experimenters including Papell and Trout (1959). A footnote to Fig. 26 states:

  > *Papell's data are for a large temperature difference between the free stream and injected air and have been corrected by* $(T_c/T_m)^{1/2}$ *as suggested by Papell to agree with present results.*

  The need for this correction indicates that η is NOT proportional to the difference between mainstream and coolant temperatures. It is not the data which are corrected. It is the *correlation* which is corrected. The correction is required because the correlation describes proportional behavior, and the data describe something quite different.

- Eq. (29-5) is the correlation actually suggested by the data--the experimental correlation:

$$T_w = 70 + (T_c - 70)/(1 + .412x) \qquad (29\text{-}5)$$

  This correlation closely agrees with *ALL* the reported data in the subject experiment and is the correlation conclusion which truly results from the experiment. In the new heat transfer, Eq. (29-5) would be presented with a description of the range of reliable application.

  If this correlation were extended by deduction in order to generate a design correlation, the area of extension would be clearly delineated, and the correlation would be accompanied by a caution that it is in large part the result of deduction and consequently is not reliable.

We therefore conclude that, even though Eqs.(29-6), (29-7), and (30-1) are generalized correlations which agree with all or some of the underlying data, there is no good reason to suppose that they predict η in practical situations with reasonable accuracy. We reach this conclusion because:

- These three correlations are largely the result of deduction rather than experiment.

- These three correlations are based on an experiment of very limited scope.

- These three correlations do not agree with other experiments of wider scope with regard to the functionality of m and $(T_m - T_c)$.

In fact, we conclude there is impelling reason to suppose that Eqs.(29-6), (29-7), and (30-1) do *NOT* predict η in practical situations with reasonable accuracy.

SCATTER

In every experiment, it is important to address the magnitude and causes of scatter. There are two types of scatter which generally concern us:

Scatter inherent in the data.

Scatter about the proposed correlation.

The scatter inherent in the data is the result of uncontrolled variables within the experiment, and imprecision in the measurements. Its magnitude can be determined by measuring the scatter in repeat data.

The scatter about a proposed correlation includes scatter inherent in the data *plus* inaccuracy in the correlation. If the scatter about the correlation equals the scatter inherent in the data, the correlation is "perfect"--ie it is the best correlation which can be induced from the given set of data.

On the other hand, if the scatter about the correlation is considerably larger than the scatter inherent in the data, we conclude that the correlation suffers from one or both of the following faults:

> The correlation includes all the important parameters, but describes the wrong functionality.
>
> The correlation omits one or more of the important parameters.

If we suspect the first fault, we will attempt to decrease the scatter by trying different forms of the correlation. If we suspect the second fault, we will attempt to decrease the scatter by finding important parameters other than those already considered.

SCATTER CONCLUSIONS--OLD WAY

In addition to their analytical correlation which is valid only in the "far from slot" regime, Hartnett, Birkebak, and Eckert (1961) present a graphical correlation which is valid in the "far from slot" regime and also in the "near to slot" regime. The graphical correlation is their Fig. 26 which describes $\eta\{x/ms\}$ on log log coordinates.

Figure 26 contains a curve which describes the authors' experimental results and is the proposed graphical correlation. The figure also contains points which describe the experimental results of other experimenters. The results plotted were obtained with m values generally in the range 0.2 to 0.4, except that the plotted results by Wieghardt (1943) contain m values in the range 0.3 to 1.0.

With regard to Fig. 26, the authors state:

> The effectiveness values reported by these (other) investigators is shown in Fig. 26 as a function of the important parameter x/ms. . . A single curve can be drawn which will represent the reported values with a deviation of ±40 per cent. *It is clearly not possible at the present time to state the specific causes of this variation . . . .*

In the "Author's Closure", the authors state:

> The differences among the values obtained in the various investigations (referring to the scatter in Fig. 26) are significant and *the authors feel that a determination of the parameters which cause these differences is one of the most urgent tasks in film-cooling research.*

In other words, the authors have investigated the scatter about the proposed graphical correlation and have correctly concluded that it is considerably greater than the scatter inherent in the data. Since the proposed correlation includes the parameters x, m, and s, the authors attribute the large scatter to the fact that parameters other than x, m, and s are important in film cooling.

They further conclude it is important to determine these other parameters which influence film cooling behavior. In fact, they feel it is

> *one of the most urgent tasks in film-cooling research.*

SCATTER CONCLUSIONS--NEW WAY

Using the new methodology, we reach the following conclusions about scatter in the results reported by Hartnett, Birkebak, and Eckert (1961):

- The scatter inherent in the data was quite small, as evidenced by the moderate scatter in Fig. (24) which is η vs. x/ms on log log coordinates, and contains results obtained at several levels of T_c.

  Figure (24) seems to suggest that the scatter about the correlation is also small--ie the figure seems to suggest that it describes the goodness of $\eta\{x/ms\}$ correlation. But the figure is misleading. Since the experiment included no variation in m or s, the figure actually makes no statement about the goodness of $\eta\{x/ms\}$ correlation. Therefore it is misleading to present the results in the form $\eta\{x/ms\}$.

- In order to present the Fig. 24 results in the proper light, they should be plotted in the form $\eta\{x\}$. And since the experimental results indicated that $\eta\{x\}$ is not a power law, there is no good reason to plot the results on log log coordinates. The results would be better shown on linear coordinates.

  In the $\eta\{x\}$ form, the figure would appropriately indicate that it does *NOT* describe the goodness of the $\eta\{x/ms\}$ correlation. It describes how well the results repeated when the coolant temperature was varied over a narrow range, and it demonstrates that film effectiveness is in fact proportional to the temperature difference between coolant and mainstream, over the small range in the experiment.

- The goodness of correlation is better described in the authors' Fig. 26 which is $\eta\{x/ms\}$ on log log coordinates. This figure includes variation in m and s by including the results of other experimenters. Fig. 26 indicates that the goodness of the $\eta\{x/ms\}$ correlation is not very good, even though m and s were varied over a narrow range--narrow compared to the range of practical application.

  Since the x axis of Fig. 26 contains the parameters x, m, and s, the figure seems to suggest that the effects of x, m, and s are accounted for in the figure. It therefore seems to support the conclusion that additional parameters are important, and must be included along with x, m, and s in order to obtain good correlation.

  In other words, Fig. 26 seems to account for the importance of x, m, and s, and seems to suggest that these parameters are not sufficient to correlate the results--seems to suggest that one or more additional parameters must be identified and included in the correlation.

- The new heat transfer has no use for contrived parameters such as m. It insists on fundamental, independent parameters, and m is a group parameter. When m is replaced by fundamental parameters, it becomes apparent that m is inversely proportional to s. And when it is recognized that m is inversely proportional to s, it becomes obvious that

  the product ms in no way depends on m or s.

Since the product ms in no way depends on m or s, it is obvious that x/ms in no way depends on m or s. And therefore correlations of the form η(x/ms) implicitly state that η *in no way depends on m or s*.

Since Fig. 26 is of the form η(x/ms), it follows that the proposed correlation in Fig. 26 implicitly states that m and s have *NO EFFECT ON FILM COOLING*, though it seems to state that m and s have first order importance!

- The end result is that we have deduced the cause of the scatter referred to in the authors' statements:

> It is clearly *not possible at the present time* to state the specific causes of this variation . . .
>
> . . . the authors feel that a determination of the parameters which cause these differences *is one of the most urgent tasks in film-cooling research*.

We have deduced that the scatter in Fig. 26 results in large part because the effects of m and s are ignored in η(x/ms) correlations such as Fig. 26. Therefore their effect on η contributes to the scatter in Fig. 26 !!

It should be noted that the methodology of the new heat transfer has made it possible to discover the cause of the scatter in Fig. 26, a discovery which is "one of the most urgent tasks in film-cooling research". Moreover, this discovery was simple--*using the new heat transfer*.

It is important to recognize that *x/ms in no way depends on m or s*. This is a very subtle point, but it must be clearly understood in order to recognize the accuracy and significance of the following observation:

> *For almost 50 years, film cooling data have been widely correlated in the form η(x/ms), a form which seems to state that m and s are of first order importance. But in fact this form states that m and s are of no importance. Even though it is widely recognized that m is of considerable importance, and s is of some importance.*

It would be *identically impossible* to devise a less effective, less desirable methodology than one which results in a correlation which seems to account for parameters it does not account for, and which should be accounted for because they greatly affect the subject phenomenon.

CONCLUDING REMARKS

I have considerable first hand, high tech experience in film cooling design/analysis/experiment/correlation. I have seen the confusion and less-than-optimum designs which result when a designer or analyst is not consciously aware that *the value of x/ms in no way depends on m or s.*

Whenever I encounter a designer/analyst whose work involves film cooling, I go out of my way to explain that x/ms in no way depends on m or s. I am often met with the disdain and lack of understanding found in Silver's (1974) review of the first edition of this book:

> (Adiutori) considers (x/ms) fallacious because he says it makes the result appear to depend on the parameters s and m whereas (xG_m/W_c') shows that s and m are unimportant, and that what matters is G_m/W_c'. This is of course nonsense. What x/ms states is a dependence on m if s is kept constant, and a dependence on s if m is kept constant . . .

The fact that x/ms in no way depends on m or s is of vital importance in the field of film cooling. Unless one has a firm grasp of this point, he/she is not truly qualified to deal with today's film cooling correlations based on x/ms.

Today's film cooling correlations are often written in terms of the group parameter ξ rather than x/ms. By definition, ξ is x/ms times the group parameter known as "slot Reynolds number". The contrived parameter ξ is worthy of note because, like x/ms, *it in no way depends on m or s!!* The reason for this total lack of dependence on m and s is that both x/ms AND slot Reynolds number *in no way depend on m or s!!!*

The difficulty with x/ms and with "slot Reynolds number" and with ξ is that they are written in terms of parameters which are *dependent*, and that is why they seem to depend on parameters on which they do not depend. This difficulty is not experienced in the new heat transfer because correlations are written in terms of parameters which are *independent*.

CHAPTER 31

DESIGN CORRELATIONS AND DOCUMENTATION

In today's old heat transfer, design correlations are generated and documented in a way which fails to rigorously establish functionality among fundamental parameters. The lack of rigor results because the experimental determination of functionality with regard to a specific parameter *need not* include variation in that parameter. The group parameters in the old heat transfer make it possible to substitute variation in one parameter for variation in another.

For example, in order to determine the effect of pipe diameter (d, in.) on the heat transfer coefficient, it is not necessary to vary pipe diameter in the experiment. The entire experiment can be performed with only one pipe!

We "know" from dimensional analysis that this type heat transfer is of the form Nu{Re,Pr}, and therefore we can determine the effect of d on h by determining the effect of Re on Nu. And since we can vary Re by varying the flow rate (W, pps), we can "determine" the effect of d on h by varying W instead of d!! And this is much more convenient than buying/fabricating/instrumenting/installing/testing several pipe test sections.

But in the new heat transfer, we consider that the effect of Parameter A can be experimentally determined ONLY by varying Parameter A in an experiment. If Parameter B is varied instead, then the experiment establishes the effect of Parameter B, but the experiment tells us nothing about the effect of Parameter A.

In the new heat transfer, *there are no group parameters*. No h; no Nu; no f; no Re; no Gr. The new heat transfer contains only fundamental parameters--q, T, W, P, E, V, I.

Since there are no group parameters in the new heat transfer, it is not possible to substitute variation in one parameter for variation in another. Experiments must address the functionality of fundamental parameters on an *individual* basis. And the fundamental parameters must retain their individual identity in design correlations and documentation. The end result is:

> Experimental correlations are necessarily generated and documented in a way which rigorously establishes the functionality among fundamental parameters *because there is no other type of parameter.*
>
> Design correlations are necessarily written in terms of fundamental parameters *because there is no other type of parameter.*

In this chapter, we discuss the old and new methodology of generating and documenting design correlations. We again use a film cooling experiment as the vehicle for illustration, and we again rely on Hartnett, Birkebak, and Eckert (1961) to explain today's version of old heat transfer methodology.

DIMENSIONLESS GROUPS

In the old heat transfer, dimensionless groups simplify experiment in two ways:

- They make it possible to obtain generalized correlations without varying each fundamental parameter of importance. It does not matter which parameter within a dimensionless group is varied, and therefore we are free to vary those parameters which are most easily varied.

  For example, suppose Tr is the dimensionless group of importance, and Tr=(XY/Z). We can vary Tr by varying X and holding Y and Z constant. In this way, we can experimentally determine the effect of both Y and Z even though *neither* is varied in the experiment.

- Dimensionless groups make it possible to perform experiments at conditions which little resemble the conditions at which the equipment will actually operate.

For example, suppose that Tr refers to the dimensionless group XY/Z, and that at the design point, X=10000, Y=75, and Z=5. If it is convenient to perform the test at X=100, we are free to do so provided we pick values of Y and Z which result in Tr=10000(75/5).

In the new heat transfer, we have no group parameters such as Nu and Re. We have only singular parameters such as heat flux, flow rate, pressure. We often identify groups of fundamental parameters which correlate experimental results, but each fundamental parameter retains its identity, and the group never attains an identity of its own.

For example, the group XY/Z might appear in a new way experimental correlation, in which case it would indicate that X, Y, and Z had each been varied over a wide range in the underlying experiment. The group parameter "Tr" would not appear anywhere because *there are no group parameters in the new heat transfer.*

Time/money constraints oftentimes make it necessary to perform an experiment without varying one or more parameters known to be important. In this event, the design correlation is obtained by extending the experimental correlation on the basis of deduction. The deduced areas of the design correlation are carefully delineated, and the prospective user is cautioned that the correlation is unreliable in the deduced areas.

FUNCTIONALITY

In today's old heat transfer, the overall purpose of experiment is to establish the functionality among dimensionless groups. This also establishes the functionality among the fundamental parameters from which the dimensionless groups are contrived.

For example, in film cooling, we might a priori deduce that the important dimensionless groups are η and x/ms. We would then establish the functionality among these groups by experiment, and we would conclude that this also establishes the functionality among x, s, mainstream flow rate, film coolant flow rate, wall temperature, mainstream temperature, and film coolant temperature.

It is generally felt that this methodology results in a reliable correlation, even if the dimensionless groups were varied by varying only one or two fundamental parameters--even if "the important parameter x/ms" was varied by varying only x, and holding m and s at fixed values throughout the experiment.

In the new heat transfer, we take a different view of this old methodology. We consider that it leads to a reliable correlation only if the experiment includes wide variation in each fundamental parameter felt to be important. If the experiment addresses film cooling, the correlation is considered generally reliable only if the experiment included wide variation in x and s and *each* fundamental parameter contained in the dimensionless groups η and m.

We note that if the experiment involved varying x while m and s were fixed, then the experiment established nothing about the functionality of m and s. Therefore it would be misleading to demonstrate that the results of such an experiment are highly correlated in the form $\eta\{x/ms\}$. The results should be used to demonstrate correlation in the form $\eta\{x\}$, since this is in fact what is demonstrated by the experiment.

In the new heat transfer, we simply do not have group parameters, and therefore it is not possible to substitute variation in one parameter for variation in another. If we wish to experimentally determine the effect of coolant flow rate on film cooling, we must actually vary coolant flow rate. We cannot substitute variation in x for variation in coolant flow rate, no matter what is indicated by a priori deduction.

In summary:

- Using the methodology of the old heat transfer, the experimental determination of functionality with regard to a specific parameter *need not include* varying that parameter in the experiment.

• Using the methodology of the new heat transfer, the experimental determination of functionality with regard to a specific parameter *necessarily includes* varying that parameter in the experiment.

THE IMPORTANCE OF FUNCTIONALITY IN DESIGN CORRELATIONS

Design correlations are used for two related purposes:

To predict equipment performance.

To optimize equipment design.

Accurate prediction of equipment performance requires design correlations which accurately describe magnitude. Design optimization primarily requires design correlations which accurately describe functionality--the functionality of fundamental parameters--*NOT* the functionality of group parameters.

For example, in film cooled equipment, "optimum" design generally refers to minimum coolant flow rate. Therefore, a design which is truly optimum must be based on a design correlation which accurately describes the functionality between film cooling effect and film coolant flow rate.

In order to demonstrate that a particular correlation accurately describes this functionality, it would be necessary to demonstrate that the value of T_w(coolant flow rate) determined from the correlation closely agrees with the value measured in the experiment. And this would require that the coolant flow rate be varied in the experiment.

This type of correlation verification little resembles the verification generally presented in the literature on the old heat transfer. Old way verification is concerned only with demonstrating agreement in terms of dimensionless groups. For example, in film cooling, verification would be demonstrated in terms of η(x/ms). Such a demonstration has little or nothing to do with verifying correlation functionality with regard to coolant flow rate.

The point to be made is that old heat transfer methodology is little concerned with the functionality of fundamental parameters--it is concerned with the functionality of group parameters. For that reason, the correlations of the old heat transfer do not properly address design optimization, even though it is one of the two main purposes of design correlations.

On the other hand, the new heat transfer methodology is concerned with fundamental parameters. In fact, it cannot avoid them *because there is no other type of parameter.* The end result is that:

- New way correlations are more likely to provide sound bases for design because they are based on experiments which directly address the functionality of designer controlled parameters.

- New way verification of experimental correlations demonstrates that they provide sound bases for design optimization because verification addresses the functionality of designer controlled parameters.

It is important to note that these desirable effects are the direct result of *abandoning* group parameters and replacing them with the fundamental parameters from which they are contrived--the parameters of real concern to equipment designers and analysts.

DESIGN CORRELATIONS--OLD WAY

Hartnett, Birkebak, and Eckert (1961) propose a graphical film cooling design correlation in their Fig. 26. With regard to film cooling design in general and the correlation in Fig. 26 in particular, the authors state:

> . . . film-cooling has already found application in gas turbines and rocket nozzles and . . .
>
> (Until better correlations are available), a reasonable estimate of the effectiveness in practical applications can be obtained from Fig. 26.

The curve in Fig. 26 describes $\eta\{x/ms\}$ results obtained at one value of mainstream flow rate, one value of coolant flow rate, one value of mainstream temperature, one value of slot height, one slot geometry, a narrow range of coolant temperatures hotter than the mainstream, and a wide range of x values.

DESIGN CORRELATIONS--NEW WAY

If the authors' Fig. 26 correlation were presented in the new heat transfer, the figure would be unchanged except that the x axis label would be replaced by

$$xG_m/W_c'$$

It would be necessary to replace x/ms because the new heat transfer has no use for group parameters such as m, and because m is dependent on s. Notice that the three parameters in the new way correlating parameter are in fact fundamental *AND* independent.

The new way description of the applicability of the graphical correlation in Fig. 26 would be:

> (Until better correlations are available), a reasonable estimate of the effectiveness in practical applications can be obtained from Fig. 26 *IF you agree with the following deductive conclusions on which this correlation is based:*

- Film cooling effectiveness is independent of G_c/G_m.

- Film cooling effectiveness is independent of s.

- The effect of mainstream flow rate can be determined by varying x rather than mainstream flow rate. (The experiment did not include variation in mainstream flow rate.)

- The effect of coolant flow rate can be determined by varying x rather than coolant flow rate. (The experiment did not include variation in coolant flow rate.)

- The effect of s can be determined by varying x rather than s. (The experiment did not include variation in s.)

- Mainstream temperature has no effect on film cooling.

- Film cooling is equal and opposite to film heating.

- The observed proportionality between film cooling effect and temperature difference between mainstream and coolant can be extrapolated to practical temperature differences.

And of course the point is that the correlation in the authors' Fig. 26 depends on so much questionable deduction that there is no hope it could provide "a reasonable estimate of the effectiveness for practical applications".

This new heat transfer view accords with the view expressed by Papell following the subject publication. With regard to the proposed correlation in Fig. 26, he stated:

> I would like to add a word of caution concerning application of the correlation presented in this paper. This correlation obtained using low temperature and low velocity data is strictly empirical and, as such, its use cannot be extrapolated to higher temperatures and velocities.

DOCUMENTATION OF EXPERIMENTAL CORRELATIONS

The literature on the old heat transfer contains numerous articles which describe experimental programs and propose experimental correlations written in terms of group parameters. The accuracy of the correlations is generally appraised by demonstrating agreement between correlation and experiment in terms of dimensionless groups.

The literature on the new heat transfer differs in the following respects:

- There are *NO* group parameters such as h, Nu, Re, f. There are only singular parameters such as q, T, W, P.

- Experimental correlations contain parameters in parametric form only if they were varied over a wide range in the experiment.

- Agreement between correlation and experiment is demonstrated for each parameter in the correlation. This is readily accomplished by presenting the results of each small experiment from which the functionality of individual parameters was induced.

- Agreement between correlation and experiment is demonstrated in an overall way by comparing correlation predictions against measured values.

- Log log coordinates are rarely used. Linear coordinates are used almost without exception.

- Functionality is never deduced as part of the preparation for experiment. It is always induced from the data.

CHAPTER 32

THERMAL BEHAVIOR CORRELATIONS FOR ONE PHASE CONVECTION

The texts and literature on today's old heat transfer indicate that convective heat transfer to a one phase fluid is accurately described by correlations of the following form:

$$Nu = a\, Re^b\, Pr^c \qquad \text{forced convection} \qquad (32\text{-}1)$$

$$Nu = a(Gr\, Pr)^b \qquad \text{free convection} \qquad (32\text{-}2)$$

This old form is totally opaque. It reveals nothing about anything of importance to designers and analysts--nothing about parameters found in Nature. If we are satisfied that these opaque equations accurately describe convective heat transfer behavior, they are readily transformed to the transparent form used in the new heat transfer. There is no need to repeat the underlying experiments unless we feel that Eqs. (32-1) and (32-2) do NOT accurately describe convective heat transfer behavior.

In this chapter, we consider thermal behavior correlations for one phase convection. These new correlations are in the following transparent form:

$$q = f\{W\}\, f\{T\}\, f\{geometry\}\, f\{\Delta T\} \qquad (32\text{-}3)$$

Since the old heat transfer indicates that Eqs. (32-1) and (32-2) accurately describe one phase convective heat transfer, we transform them into thermal behavior correlations. But we intend that data from future experiments will be obtained and correlated in the manner suggested by the new heat transfer. And we expect that these future experiments will indicate the need for changes in substance as well as form.

FORCED CONVECTION THERMAL BEHAVIOR CORRELATIONS

Until experiments are performed within the framework of the new heat transfer, forced convection thermal behavior correlations will be obtained by transforming the generally dimensionless correlations of the old heat transfer. They have been used so widely and for so long that it seems possible they agree reasonably well with the data.

It would be desirable to determine whether or not Eqs. (32-1) and (32-2) accurately describe the behavior of the fundamental parameters they implicitly describe. This can not be done because it would require data, and today's heat transfer literature does not contain data. It generally contains only dimensionless group parameters, and it is not possible to verify the correlation functionality of individual parameters from a specification of the "measured" values of group parameters.

The transformation of old way correlations requires that group parameters such as h, Nu, Re be replaced by the fundamental parameters from which they are contrived, such as q, T, W. (The transformation details are described in Chapter 9.) Eq.(32-4) is the result of transforming Eq.(32-1) from a fluid general, dimensionless Nusselt number correlation to a fluid specific, dimensional thermal behavior correlation--the preferred form in the new heat transfer:

$$q = a_1 W^b x^{b-1} f\{T\} \Delta T \qquad (32\text{-}4)$$

$$f\{T\} = k^{(1-c)} (C/\mu)^c$$

Transformed correlations are intended to be used only until thermal behavior correlations have been obtained using the experimental methodology of the new heat transfer. (The new experimental methodology is described in depth in the first edition of this book, and in summary in Chapters 23 to 32 of this edition.)

FREE CONVECTION THERMAL BEHAVIOR CORRELATIONS

The transformation of Eq.(32-2) to a thermal behavior correlation results in Eq.(32-5).

$$q = a_2\ x^{(3b-1)}\ f\{T\}\ \Delta T^{(1+b)} \tag{32-5}$$

$$f\{T\} = k^{(1-b)}(\rho^2\beta C/\mu)^b$$

Chapter 9 describes the details involved in transforming analytical correlations. However, free convection correlations are sometimes presented graphically, and their transformation is somewhat more involved. It is accomplished in the following steps:

- Obtain a set of Nu{GrPr} coordinates by inspection of the given graphical correlation.
- At each pair of coordinates, calculate NuGrPr.
- Plot NuGrPr vs GrPr.
- Label the NuGrPr axis $qx^4\rho^2 g\beta C/\mu k^2$.
- Label the GrPr axis $\Delta T\rho x^3 C/\mu k$.

The transformed graph is in thermal behavior form because the y axis contains q and NOT ΔT, and the x axis contains ΔT and NOT q. In other words, q and ΔT are separate in the graph.

THE PROBLEM WITH FREE CONVECTION h CORRELATIONS

Notice that q and ΔT are *NOT* separate in Nu(GrPr) graphical correlations because ΔT is implicit in BOTH Nu and GrPr. For this reason, Nu(GrPr) graphs are not easy to read if q is known and ΔT is sought. When this is the case, an iterative technique is required *simply to read the graph!*

This can be seen by noting that, if q is given and ΔT is desired, it is not possible to calculate GrPr because we don't know ΔT, and it is not possible to calculate Nu because we don't know $q/\Delta T$ (alias h). Therefore the solution requires an iterative technique in which we:

- Guess at a value of GrPr.

- Read the graph to obtain a first estimate of Nu.

- Use the estimate of Nu to estimate the value of ΔT.

- Use the estimate of ΔT to determine a better estimate of GrPr.

- Continue iterating until the solution converges on the desired value of ΔT.

This procedure is time consuming at best, and texts on the old heat transfer generally avoid discussing it. Textbook problems which illustrate the use of Nu(GrPr) graphs are generally stated with ΔT known and the value of q unknown. In this case, Nu(GrPr) graphs can be read directly because GrPr contains ΔT and not q.

THE IMPORTANCE OF SEPARATING THE VARIABLES

The fact that Nu{GrPr} graphs must often be read using an iterative technique is the *direct result* of combining q and ΔT in Nu. (Nu is usually written hD/k, but it is really $qD/\Delta Tk$ because h is really $q/\Delta T$.)

It is important to note that the new way graphical correlation can be read directly whether q is known and ΔT is sought, or ΔT is known and q is sought. The ability to read the graph directly in both directions is the direct result of keeping q and ΔT *separate*.

Notice that we are dealing here with moderate nonlinearity in $q\{\Delta T\}$, and even moderate nonlinearity causes the difficulty of solution to greatly depend on whether we deal with combined variables or separated variables--whether we deal with $q/\Delta T$ (alias h) or $q\{\Delta T\}$ (alias thermal behavior)--whether we use the old heat transfer or the new heat transfer.

In the later chapters, we will find that as the degree of nonlinearity increases, the difference in the degree of difficulty involved in problem solution also increases. When we consider the highly nonlinear behavior exhibited by boiling phenomena, we will find that problem solution based on h is virtually impossible, whereas the solution of the same problems is quite simple *if the solution is based on* $q\{\Delta T\}$ *and the new heat transfer*.

CHAPTER 33

THE POOL BOILING CURVE--OLD WAY

In today's old heat transfer, the "pool boiling curve" (PBC) describes q{ΔT} in a vessel which contains saturated liquid, and through which there is no net flow. The PBC aids in understanding and describing the behavior of pool boiling equipment. It does not replace "h" in the design and analysis of pool boiling equipment.

Because of its highly nonlinear character, the PBC is normally broken up into regimes. Analytical design correlations are generated for each regime in the h{ΔT} form required for design and analysis in the old heat transfer. Correlations are also generated in the q{ΔT} form; they are readily transformed to h{ΔT} by dividing q by ΔT.

This chapter presents an overview of the current, widely accepted version of the PBC. It notes that:

- The old PBC contains a region in which boiling does *NOT* occur.
- The q{ΔT} form is used in order to isolate errors in the data which would be less apparent in h{ΔT} form.
- The logarithmic coordinate system used for PBC's is apparently the result of assuming that boiling regimes are described by power laws.

This chapter is preparation for a later chapter in which we discuss the PBC in the new heat transfer. The new PBC differs from the old in both functionality and application. It contains several regions which are quite linear, and it is primarily a graphical correlation used to design and analyze pool boiling equipment.

THE GENERIC POOL BOILING CURVE

In the old heat transfer, the generic "pool boiling curve" is sometimes given in the form log h vs log ΔT as in Holman (1986), but it is usually given in the form log q vs log ΔT as in Kreith and Bohn (1986) and in Incropera and Dewitt (1985). The old generic pool boiling curve resembles Fig. (33-1) :

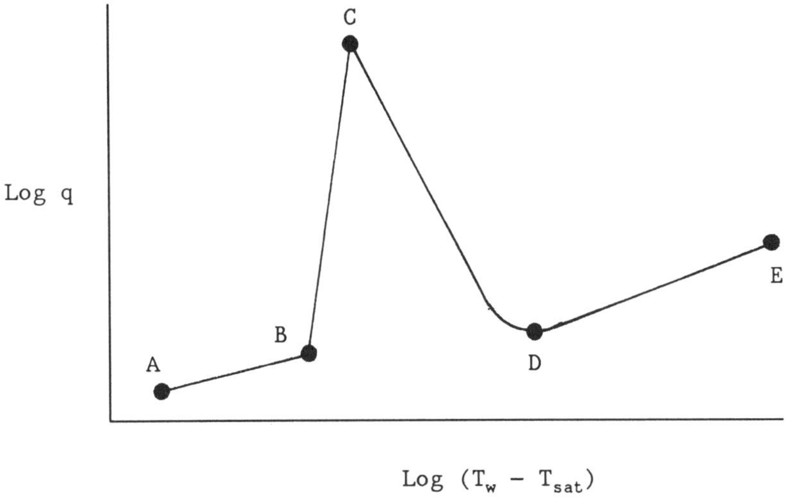

FIGURE (33-1) Typical pool boiling curve (old way)

Notice the following in Fig. (33-1):

- The PBC is in the form $q\{\Delta T\}$, and this is *NOT* the form required for design and analysis in the old heat transfer.
- The PBC contains a sharp maximum at C and a blunt minimum at D.
- The pool boiling curve includes a region in which no boiling occurs--the free convection regime, region AB.
- The pool boiling curve is presented on log log paper. Therefore the straight line segments are power laws.

THE REGIMES

The PBC is generally broken up into at least four regimes, as shown in Fig. (33-1). Experiments do not usually deal with all four regimes, and correlations generally deal with only one regime--ie correlations are regime specific. In order of increasing ΔT, the regimes are:

Free convection regime AB:

- The surface temperature exceeds saturation temperature, but the difference in temperature is not large enough to generate bubbles.

- *There is no boiling.* Heat transfers from the heated surface to the free surface by one phase free convection.

- Heat transfer coefficients are calculated from one phase, free convection correlations which indicate that $q\{\Delta T\}$ is a power law with an exponent of 5/4 or 4/3.

Nucleate boiling regime BC:

- The surface temperature is greater than the temperature required to generate bubbles, but less than the temperature at the maximum in $q\{\Delta T\}$.

- The Rohsenow (1952) $q\{\Delta T\}$ correlation is widely referenced. It is used directly to solve simple problems which involve only the boiling interface.

- If the problem involves a boiling interface in series with a wall and another interface, the Rohsenow correlation cannot be used directly because it is in the $q\{\Delta T\}$ form. It must first be transformed to the $h\{\Delta T\}$ form generally required in the old heat transfer.

- It is widely agreed that $q\{\Delta T\}$ is a power law with exponent 3 or 4. (The exponent in the Rohsenow correlation is 3.)

- Surface condition and fluid-surface combination have a strong influence on $q\{\Delta T\}$.

Transition boiling regime CD:

- Surface temperature is between the maximum and the minimum in $q\{\Delta T\}$.

- There is no widely accepted analytical correlation for q or h.

- Berenson (1962) states: "It was found, with the exception of some of the data . . . that the transition boiling data lie along a straight line connecting the burnout point and the film-boiling minimum point on log-log graph paper. This is also true of the transition boiling data obtained by Braunlich (1941) and Kaulakis and Sherman (1938)."

- Addition of wetting agents to boiling liquid has a strong influence on $q\{\Delta T\}$.

Film boiling regime DE:

- Surface temperature is greater than the temperature at the minimum in $q\{\Delta T\}$.

- Kreith and Bohn (1986) cite film boiling $h\{\Delta T\}$ correlations by Bromley (1950) and Westwater and Breen (1962).

THE REASON FOR $q\{\Delta T\}$

The rationale behind the use of $q\{\Delta T\}$ rather than $h\{\Delta T\}$ is described in a comment made by McAdams in response to a paper by Farber and Scorah (1948). The paper contained a number of PBC's in the form $h\{\Delta T\}$. In the discussion following the presentation, McAdams referred to the $h\{\Delta T\}$ PBC's in the following:

> *Because of the possibility of error in temperature measurement, it would be safer to plot heat-flux density versus temperature, rather than coefficient versus temperature, in order to isolate any possible error in the abscissa.*

Notice that McAdams' comment has nothing to do with the design or analysis of equipment. It is concerned only with identifying experimental error.

The old heat transfer employs the form q{ΔT} to present and compare pool boiling data, but it is not intended to replace h in the design/analysis of pool boiling equipment. Pool boiling publications oftentimes present the results in the h form required for design/analysis, and also in the q{ΔT} form required "to isolate any possible error in the abscissa".

For example, Eckert and Drake (1972) present a PBC and a comprehensive discussion of boiling heat transfer coefficients. Cooper (1984) discusses nucleate pool boiling q{ΔT} curves, and presents an in-depth analysis of nucleate pool boiling heat transfer coefficients. Thome and Shock (1984) discuss pool boiling q{ΔT} curves and pool boiling heat transfer coefficients.

THE REASON FOR LOGARITHMIC COORDINATES

The pool boiling curve is presented on log log coordinates either because the data generally describe power laws, or because power laws are simply assumed. Since much of the old heat transfer is based on the assumption of power law behavior, it seems likely that the coordinate system is the result of assumption. (In any event, we should and will review the data in the literature in order to independently determine the best coordinate system to be used in the new heat transfer).

THE REASON FOR A NON-BOILING REGION

It is self-evident that a "pool boiling curve" should describe boiling behavior, and yet the old PBC includes a non-boiling region. This anomaly results because, in the old heat transfer, the PBC is NOT a description of boil*ING* behavior--it is a description of boil*ER* behavior. For this reason, "pool boil*ER* curve" would be a more appropriate name for the PBC of the old heat transfer.

The generation of bubbles in a liquid phase requires a finite temperature difference. Boiling generally does not occur until the heated surface exceeds the saturation temperature by a considerable amount. The end result is that *no boiling occurs at the lower end of the PBC!* In that region, the heat transfer mechanism is free convection at the solid/liquid interface, and evaporation at the liquid/vapor interface.

(Boiling occurs at all points of the new "pool boiling curve" because it is a description of boil*ING* behavior.)

THE MAXIMUM AND MINIMUM HEAT FLUX

The heat flux at the maximum in $q\{\Delta T\}$ is often referred to as "the maximum heat flux". In many applications, the maximum heat flux is by far the most important parameter in the PBC because operation at the maximum endangers the equipment. In such applications, the equipment is usually designed to ensure operation well below the maximum at all times.

Because of the considerable practical importance of the maximum heat flux, the literature contains a great many correlations which describe how system parameters affect the maximum heat flux.

The literature also contains correlations which describe how system parameters affect the heat flux at the minimum in $q\{\Delta T\}$--the "minimum heat flux". These do not have the importance of the correlations which describe the maximum.

CHAPTER 34

NUKIYAMA'S PIONEEERING EXPERIMENT

Nukiyama (1934) is generally regarded as the pioneer of the pool boiling curve. His work marks the beginning of a real understanding of boiling phenomena. To honor his important contribution, the International Journal of Heat and Mass Transfer published an English translation of his pioneering article in 1966, and again on the 50th anniversary of its original publication in Japan.

The 1984 re-publication includes an introduction entitled "Memories of My Research on Boiling" written in 1984. In it, Nukiyama describes the status of boiling research in 1934:

> Old literature had described the q vs T characteristic (at only a fraction of the maximum heat flux) because it had been assumed that heat transfer would be through a flat plate or metal tube. q increased with T and no way had been found to supply enough q to raise T further.

In other words, no experimenter had been able to reach the maximum in the pool boiling curve in a steady-state way because the heating methods used were not adequate to achieve the required heat flux level. Pool boiling test equipment had always been designed with a heated surface which was either a flat plate or a tube.

Nukiyama describes the experimental method he devised in order to attain higher levels of heat flux than had previously been attained in boiling experiments:

> I devised a new experimental method where a metal wire was electrically heated in the water. The heat transfer surface area was assumed to be the surface area of the metal wire, and q and the wire temperature were calculated from the electric current and resistance . . .

He also devised/invented a new type of flat plate pool boiler in which the heated region was the upper end of a heat flux amplifier. The amplification made it possible, for the first time, to operate a flat plate boiler at the maximum in $q\{\Delta T\}$.

Nukiyama's experiment was incredibly comprehensive, particularly for a pioneering effort, and included all of the following:

- Measurement of the pool boiling curves at heat fluxes to the maximum in $q\{\Delta T\}$ using platinum wires, nickel wires (2 diameters), and nichrome wires (2 diameters). 54 runs were made.

- Measurement of the pool boiling curves at heat fluxes to the maximum in $q\{\Delta T\}$ using flat copper plates (2 diameters). 7 runs were made.

- Measurement of the pool boiling curve at the minimum in $q\{\Delta T\}$ and 1000 C beyond the minimum using platinum wires. 12 runs were made.

- Measurement of the pool boiling curve at heat fluxes to the maximum in $q\{\Delta T\}$ with a hollow, unheated, vertical cylinder sitting on the heated flat plate surface. The hollow cylinder contained a guillotine divider which could be adjusted vertically, and the effect of divider location on $q\{\Delta T\}$ was also measured.

- All the above runs were performed with saturated water in the boiler. Three additional runs were performed to measure the effect of subcooling on the pool boiling curve at heat fluxes to the maximum in $q\{\Delta T\}$.

NUKIYAMA'S RESULTS

Nukiyama's results are described by the following:

- The results are presented as $q\{\Delta T\}$ and $h\{\Delta T\}$.

- The results are presented on *LINEAR* coordinates.

- There is a maximum in $q\{\Delta T\}$ at approximately 50 C.

- There is a minimum in q{ΔT} at approximately 300 C.

- No data is presented in the region between the maximum and the minimum in q{ΔT}.

- The region below the maximum in q{ΔT} contains a knee at approximately 10 C. Presumably the knee separates the non-boiling, free convection regime and the fully developed nucleate boiling regime.

- The nucleate boiling region above the knee does *NOT* suggest power law behavior. Above the knee, the curves describe lines which are reasonably straight. In other words, the data indicate that the thermal behavior of nucleate boiling is more or less *linear*.

- In the nucleate boiling regime, the slope of the curves sometimes decreases with temperature and sometimes increases, but not enough to suggest correlating the data in a nonlinear way.

- If the nucleate boiling data described power laws, the slope of the curves would generally INcrease with temperature (since the second derivative of a power law is positive if the exponent is 3 or 4). The slope of the curves in Nukiyama's article DEcreases more generally than it increases.

- The film boiling curves do not suggest power laws because the extrapolation of the curves does not pass through the origin as would power laws. (The curves contain data in the wide range 200 to 1200 C, and therefore only a 10% extrapolation is required.) This may reflect that all of the wire was not in the film boiling regime.

- Subcooling has a positive impact on the heat flux at the maximum in q{ΔT}, and a negative impact on the surface temperature at a given heat flux.

- A hollow, unheated, vertical cylinder sitting on a flat, heated surface has a major and positive impact on the heat flux at the maximum in q{ΔT}. The impact on surface temperature at a given heat flux is secondary.

- A divider in a hollow cylinder has a major and positive impact on the maximum heat flux. The effect depends on divider location, and there is an optimum location in the central region of the cylinder.

Notice in particular that the measured functionality in nucleate boiling little resembles the power laws used to describe nucleate boiling in today's old heat transfer.

A COMMON MISCONCEPTION ABOUT BOILING

Nukiyama's reminiscence includes a candid description of a misconception he had about boiling prior to his experiment, and which is still quite common:

> In the early stages of my study, I found that the temperature of a metal wire easily reached as high as 105 C without the water boiling. I was in the skies because this was contrary, or so I thought, to the invariable principle that "Water boils at 100 C" (a lone exception is super-heated water which is not easily attained). However, when I happened to read an old textbook, *Theory of Heat*, written by Clerk Maxwell, Lord Rayleigh and others, it was lightly described that water boiled when it reached the pertinent boiling temperature for a certain pressure plus the temperature at which the cohesion of the water and its contact surface was overcome, and I realized that they had already known the phenomenon.

Notice Nukiyama's boundless enthusiasm on "discovering" that boiling does not begin at 100 C:

> I was in the skies because this was contrary . . .

and contrast his enthusiasm with the consternation which today would greet any experimental result which seemed contrary to the widely accepted view!

CHAPTER 35

THE INCEPTION OF BOILING

Today's old pool boiling curve (PBC) indicates a smooth transition from the non-boiling, free convection regime to the nucleate boiling regime. The curve indicates that the transition involves no step changes in $q\{\Delta T\}$, and there is nothing to suggest that the transition sometimes involves undamped oscillatory behavior.

In this chapter, we examine several literature articles to determine how the transition from non-boiling to boiling actually occurs in a pool boiler. This examination indicates that the transition is *NOT* smooth. The transition sometimes involves step changes in temperature of 25 to 50 F, and sometimes results in undamped, oscillatory behavior during which the surface temperature fluctuates 150 F.

The undamped oscillatory behavior is indicative of thermal instability, and this subject is not within the knowledge envelope of the old heat transfer. For that reason, the significance of the oscillations was not recognized.

In the next chapter, we determine the significance of the oscillations with regard to the generic PBC. In a later chapter, we consider the new PBC, and we include revisions suggested by the step changes and oscillations which have often been observed at the inception of boiling.

THE EXTENT OF THE NON-BOILING REGIME

Boiling does not occur in pool boilers until the temperature of the heated surface exceeds saturation temperature by a considerable amount. The magnitude of the required superheat is indicated by Rohsenow (1973):

> When a pool of liquid at saturation temperature is heated . . . the appearance of the first bubble . . . requires a significant finite superheat. . . Some experiments have shown that at a heated surface in water at atmospheric pressure, boiling begins at around 30 F above saturation temperature.

Corty and Foust (1955) observed the pool boiling behavior of several organic liquids. Their results indicate that approximately 50 F is required to initiate pool boiling, and approximately 25 F is required to sustain it.

Nukiyama (1984), in describing his PBC research in the 1930's, states:

> In the early stages of my study, I found that the temperature of a metal wire easily reached as high as 105 C without the water boiling.

THE TRANSITION TO BOILING--ORGANIC LIQUIDS

Corty and Foust (1955) give the following detailed description of the transition from non-boiling to boiling in a pool boiler containing an organic liquid:

> When boiling occurred at a fairly high rate, (q=15000 $B/hrft^2$), the surface was densely covered with hundreds of bubble columns per square inch. As the flux was reduced, more and more of these died out (and then) the few remaining nuclei died out one by one.
>
> If the reduction of flux was interrupted and energy increased again, the stage at which this increase was started governed the way in which nucleate boiling was reestablished at the higher rates.

> If all the active centers were allowed to disappear and the surface was kept free of bubbles for at least 10 to 15 min., it was possible to retain free-convection heat transfer to much higher ΔT's than would normally be required for vigorous nucleate boiling. Superheats of 40 to 50 F above the saturation temperature of the liquids were possible with no bubbles on the heat transfer surface even though the normal ΔT for violent nucleate boiling was only about 25 F. Such excessive superheats could be maintained for several minutes, but upon further increases the surface spontaneously broke into vigorous nucleate boiling, and the ΔT then decreased to "normal" value . . .
>
> If the last remaining active nuclei were just permitted to die out and the power input was immediately increased, the nuclei last active reactivated first. Upon further increase in power these nuclei became the centers of patches on which boiling occurred violently while the rest of the surface was bare. The widely varying readings of the thermocouples indicated that these patches were at temperatures characteristic of normal nucleate boiling, although the bare spots on the surface were at much higher temperatures. . . At fixed rates of energy input the patches were stable, and they extended only when the power was raised. When flux was sufficiently great (about 40000 B/hr ft^2) the surface was again completely covered with bubbles and the hysteresis effect vanished.
>
> If the decrease in flux was interrupted while there still were a number of bubbles active, the pattern of patchwise boiling emerged again. In that case the larger number of bubble patches merged quickly to reestablish normal nucleate boiling . . .

Schnurmann and Lardge (1973) obtained similar results with organic liquids. They state:

> The transition point from (free) convection to nucleate boiling is clearly marked by a sharp drop of the temperature difference between the wire and the bulk of the liquid . . .
>
> At this transition point there may be a temperature drop of nearly 30 C with n-heptane and about 18 C with perfluoromethylcyclohexne.

In summary, the transition from non-boiling to boiling often results in a step decrease in surface temperature of 25 to 50 F when boiling organic fluids in an electrically heated pool boiler.

THE TRANSITION TO BOILING--LIQUID METALS

The transition from non-boiling to boiling is extremely non-smooth when dealing with liquid metals. Colver and Balzhiser (1965) describe the transition in an electrically heated pool boiler which contains a liquid metal:

> Temperature fluctuations up to 150 F (were observed in surface thermocouples) at flux levels below 250,000 B/hr ft^2. In some cases the temperature increased rapidly to a point, then slowly (approximately 1 F/sec) to a maximum value. At this point it instantaneously dropped to the original value. Many times this drop exceeded the initial temperature rise and fell as much as 50 F below the original level. . . Each instantaneous drop in temperature was accompanied by a pressure surge and an audible bump. The largest bulk variation . . . was 70 F. . . Above 250,000 B/hr ft^2 the amplitude of fluctuations in the boiling tube decreased significantly (and) the bulk temperature and system pressure remained essentially constant.
>
> (This behavior) indicates that . . . the nucleate boiling process . . . was periodically lost with the system reverting back to convection heat transfer. The upward spikes then may represent a superheating of the system . . . When the superheat was again great enough, the system broke into vigorous boiling with a resultant temperature drop. Many times this drop . . . approached the liquid saturation temperature. Immediately after this, the temperature fluctuations indicated stable boiling. After a time, dependent on heat flux level, stable boiling was lost and the process repeated. . . At high heat fluxes . . . fluctuations were generally not detected.

Similarly, Marto and Rohsenow (1966) describe the transition in an electrically heated pool boiler which contains a liquid metal:

> During nucleation, large boiler wall temperature fluctuations occurred which in some cases were as high as 150F.
>
> These fluctuations were always accompanied by large variations in the test section noise level as determined from the phonograph cartridge. The sharp increase in noise level and the sudden decrease in wall temperature of the boiler always occurred coincidentally . . . This is interpreted to be the onset of nucleate boiling. After this "bump", nucleation may continue . . . as evidenced by the continued noise level and lower wall superheat. . . When the noise stops, the temperature rises gradually to its maximum value.
>
> When boiling is stable, the wall temperature remains at the lower level and the noise persists.
>
> All the unstable data show that, as the heat flux is increased, stability improves . . . The experimental results show that, around 200,000 B/hr ft^2, stable boiling occurs in most cases.

Marto and Rohsenow refer to the undamped oscillatory behavior as "nucleate boiling instability", but they do not identify the cause of the oscillatory behavior, nor do they reach any conclusion about its significance with regard to the PBC.

CONCLUSIONS

The experimental observations described above suggest the following conclusions:

- The transition from non-boiling regime to nucleate boiling regime is not generally smooth.

- The ΔT which is required to initiate nucleate boiling in an organic liquid is considerably greater than the ΔT required to sustain it. The end result is a step decrease in surface temperature at the inception of boiling. The magnitude of the step is 25 to 50 F.

- The PBC for liquid metals is such that it results in undamped oscillatory behavior when the pool boiler is operated at heat flux levels below a critical value. The amplitude of the oscillations is approximately 150 F.

The oscillatory behavior is indicative of thermal instability, and suggests that boiler operation alternates between boiling and non-boiling.

As we noted in a previous chapter, an understanding of thermal stability is not within the knowledge envelope of the old heat transfer. For that reason, it was simply not possible for experimenters to determine the significance of the undamped oscillatory behavior.

In the next chapter, we determine the cause of the undamped oscillatory behavior, and also its significance with regard to the PBC for liquid metals. In a later chapter, we make a number of revisions to the old PBC, including those which are suggested by the step changes and undamped oscillations noted above.

CHAPTER 36

UNDAMPED OSCILLATORY BEHAVIOR--ANALYSIS AND PBC SIGNIFICANCE

Electrically heated pool boilers which contain liquid metals have often been observed to operate in an undamped oscillatory manner. This behavior occurs at the lower end of the nucleate boiling regime, and transient data suggests that the oscillations involve alternate periods of boiling and non-boiling.

In this chapter, we consider the underlying cause of the oscillatory behavior, and its significance with regard to the pool boiling curve (PBC). We note that oscillatory behavior is indicative of some form of instability, and we attack the problem by postulating that thermal instability is the underlying cause. (See Chapter 11.) In particular, we answer the questions:

- How could thermal instability result in undamped oscillations in an electrically heated pool boiler?
- What $q\{\Delta T\}$ would a PBC have to describe in order that thermal instability would result in undamped oscillatory behavior?
- If a PBC described the $q\{\Delta T\}$ which results in undamped oscillations, how would the boiler behave in terms of $q\{t\}$ and $\Delta T\{t\}$? ("t" is time, seconds.)

When we have answered these questions, we qualitatively compare the $q\{t\}$ and $\Delta T\{t\}$ answers with values observed and reported in the literature. We find that the agreement is excellent, and therefore we conclude that the postulated PBC shape is the actual PBC shape.

THE POSTULATED CAUSE OF THE OSCILLATIONS

Our analysis is based on postulating that thermal instability is the underlying cause of the observed oscillatory behavior. As we noted in Chapter 11, thermal instability results if, at any plane in the system,

$$(dq_{out}/dT) \leq u\ (dq_{in}/dT) \qquad (36\text{-}1)$$

With regard to a pool boiler, the left side of Criterion (36-1) is the slope of the PBC, the right side is the derivative of heat flow into the boiling interface with respect to the temperature of the boiling interface.

The boiler is electrically heated, and let us assume for simplicity that the heating elements have a temperature coefficient of resistivity equal to zero, in which case $dq_{in}/d\Delta T$ is zero. We therefore conclude that, if the boiler is thermally unstable, the q functions into and out of the boiling interface must have the relationship shown in Fig. (36-1).

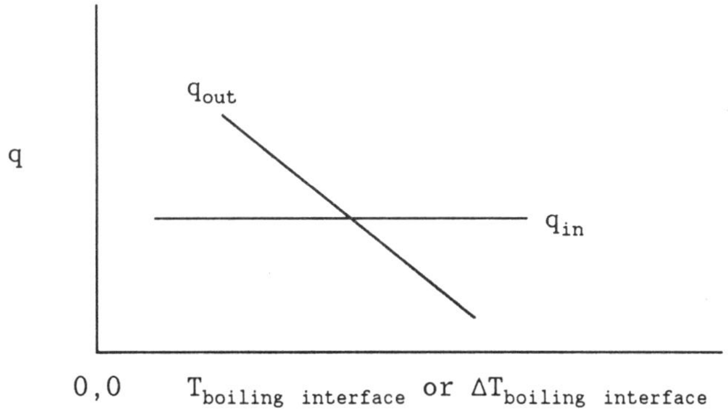

FIGURE 36-1 The Postulated Cause of Undamped Oscillations

Fig. (36-1) indicates that the heat flux into the heated wall is independent of its temperature because the boiler is electrically heated, and we have assumed that the temperature coefficient of resistivity is zero. Because we have postulated that thermal instability is the cause of the undamped oscillations, Fig. (36-1) indicates that the slope of the PBC is negative because thermal instability results only if the slope of the PBC is less than the slope of $q_{in}\{\Delta T\}$.

THE DEDUCED PBC

Undamped oscillatory behavior indicates that:

- The equipment is operating at an unstable intersection of q_{in} and q_{out}.
- Without outside influence, the equipment cannot "find" a stable intersection of q_{in} and q_{out}.

If we can deduce undamped oscillatory behavior from a postulated PBC, it will demonstrate a real understanding of the cause of the oscillations, and also will reveal something about the PBC. Therefore the problem becomes: *How should the functions in Fig.(36-1) be completed so that the equipment would be unable to "find" a stable operating point?*

Fig.(36-2) answers this question. The key point in the figure is that the negative slope region BC is bounded by endpoints at which $dq/d\Delta T$ is infinite.

> *Infinite slope at the endpoints of a negative slope region prevents the equipment from "finding" a stable operating point, and results in undamped oscillatory behavior.*

HOW TO FIND THE ENDLESS LOOP IN FIG. (36-2)

Figure (36-2) describes steady-state behavior, but it can be used to determine transient behavior during undamped oscillatory operation. This requires noting that:

- If the heat flow into the wall is greater than the heat flow out, the wall temperature increases and conversely.

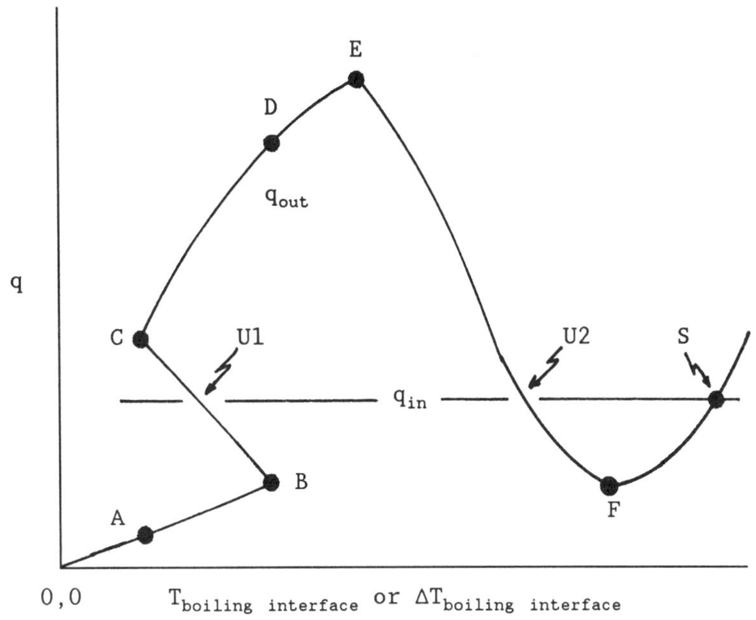

FIGURE (36-2) The Postulated q_{in} and q_{out} Functions

- If the wall temperature increases, operation moves to the right in the figure, and conversely.
- The wall temperature is continuous with time. Heat flux out of the wall is not necessarily continuous with time.

To illustrate: If the boiler is initially operating at the U1 intersection in Fig. (36-2), and it then receives a small perturbation to the right, the transient behavior which results is deduced as follows:

- Fig. (36-2) indicates that, on the right side of the intersection, the heat flow into the boiler wall exceeds the heat flow out, causing the wall temperature to increase until it reaches point B.
- At point B, operation jumps to point D.
- At point D, the heat flow out of the wall is greater than the heat flow into the wall, causing the wall temperature to decrease until it reaches point C.

- At point C, operation jumps to point A.

- At point A, the heat flow into the wall exceeds the heat flow out of the wall, causing the wall temperature to increase until it reaches point B.

The endless loop in Fig. (36-3) describes the oscillatory behavior which would result if q_{in} were anywhere in the BC region, and if its value had not exceeded the value at Point E since the last time the equipment was started up.

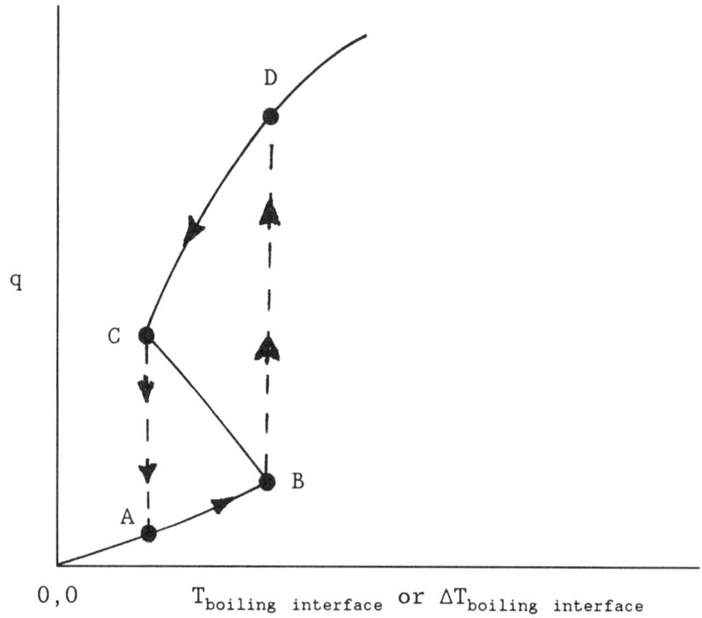

FIGURE (36-3) The Endless Loop in Fig. (36-2)

Notice that, if Fig. (36-2) had been drawn so that ΔT at the maximum was less than the value at B, the loop in Fig. (36-3) would include operation in the transition region.

Also notice that, if Fig. (36-2) had been drawn so that q at B was greater than q at F, and B was drawn to the right of EF, the boiler would not operate in either the nucleate or the transition regime! As the boiler power was slowly increased, the boiler would pass directly from the free convection regime to the film boiling regime!

This unusual behavior has been reported in the literature. For example, Winterton (1983) states:

> Lyon, Foust and Katz (1955) boiled mercury on the outside of a horizontal steel tube at 1 atm. They concluded that as the power was raised the mercury went straight from natural convection to film boiling . . . In this power controlled system it was not, of course, possible to obtain transition boiling data. . .

It is important to note that this strange behavior does NOT establish that the PBC contains no nucleate or transition boiling regions. It establishes only that, *in this particular case,* the shape of the PBC prevents this particular boiler from "finding" the nucleate and transition boiling regions.

DEDUCING q{t} AND ΔT{t} FROM FIG. (36-3)

We can qualitatively determine q{t} and ΔT{t} by inspection of Fig. (36-3). The result is shown in Fig. (36-4).

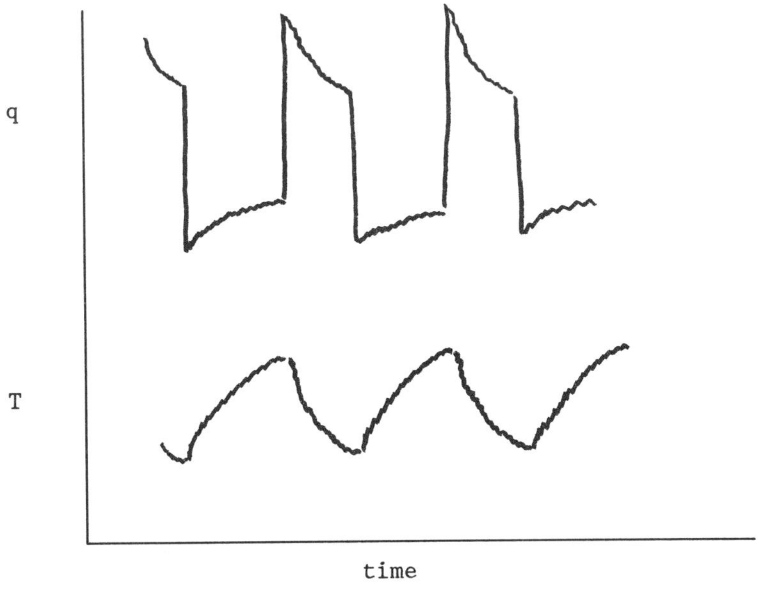

FIGURE 36-4 q{t} and T{t} from Fig. (36-3)

THE PBC CONCLUSION

The PBC conclusion is that boiling liquid metals often exhibit a PBC which resembles the curve in Fig. (36-2). This conclusion is supported by the following reasoning:

- Electrically heated pool boilers which contain liquid metals generally behave in an undamped oscillatory manner at the lower end of the nucleate boiling regime.

- The deduced PBC in Fig. (36-2) would result in undamped oscillatory behavior at the lower end of the nucleate boiling regime in a typical, electrically heated boiler.

- The undamped oscillatory behavior observed in liquid metal pool boilers alternates between boiling and non-boiling.

- The deduced PBC in Fig. (36-2) results in undamped, oscillatory behavior which alternates between boiling and non-boiling.

- The T{t} deduced from Fig. (36-2) and described in Fig. (36-4) closely resembles the T{t} observed and reported by Marto and Rohsenow (1966).

THE BEHAVIOR WITH A CONSTANT TEMPERATURE HEAT SOURCE

Even if the boiler heat source were a condensing vapor, the undamped oscillatory behavior described in Figs. (36-2) to (36-4) would result in the BC region *provided* the inequality in Crit. (36-1) were satisfied.

If the boiler design were such that Criterion (36-1) were NOT satisfied in the BC region, operation would exhibit the pronounced hysteresis described in Fig. (36-5). (We will return to this subject in a later chapter.)

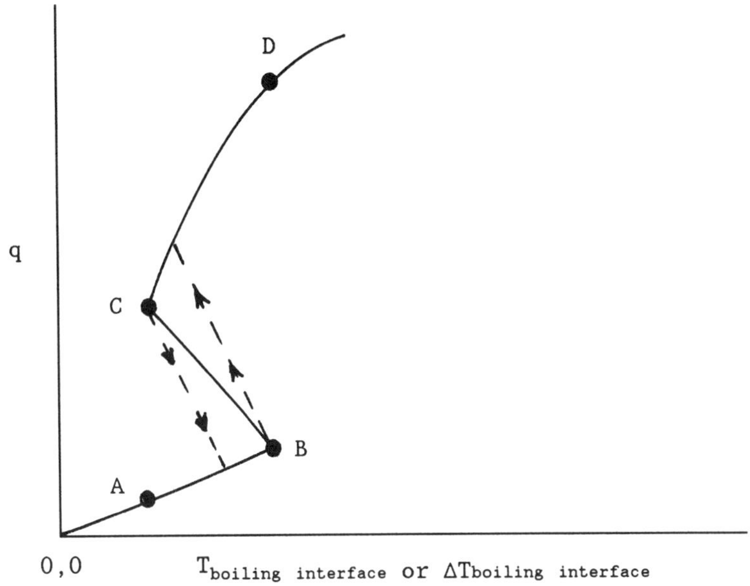

FIGURE (36-5) Hysteresis Exhibited by a Temperature Controlled Boiler--Crit. (36-1) NOT Satisfied

CHAPTER 37

NUCLEATE BOILING THERMAL BEHAVIOR AND LOG LOG PAPER

In today's old heat transfer, it is widely agreed that $q\{\Delta T\}$ in nucleate pool boiling is highly nonlinear. For 40 years, the behavior of nucleate pool boiling has been described by power laws in the form of Eq. (37-1) where the exponent n is "3 or 4".

$$q \propto \Delta T^n \qquad (37\text{-}1)$$

But it should be borne in mind that, in the old heat transfer, power laws and log log paper usually represent assumption rather than experimental conclusion. Therefore it is reasonable to consider that nucleate boiling power laws may also be the result of assumption, in which case they would be suspect.

In this chapter, we review the nucleate boiling data in the literature and the correlation methodology which results in power laws. We conclude that:

- The correlation methodology generally used in the old heat transfer is not rigorous.
- Nucleate boiling data in the literature generally indicate that nucleate boiling behavior is highly *LINEAR*.
- The literature on the old heat transfer contains several articles which rigorously correlate nucleate boiling data and conclude linearity. Unfortunately, they are seldom referenced in the literature on the old heat transfer.

NUCLEATE POOL BOILING THERMAL BEHAVIOR--OLD WAY

For 40 years, it has been widely agreed in the old heat transfer that nucleate pool boiling q{ΔT} is described by highly nonlinear power laws.

- McAdams (1954) states:

  in the range of strong nucleate boiling . . . the data of many observers may be expressed by

  $$q = a\,\Delta T^n \qquad (37\text{-}1)$$

  where n is a constant ranging from 3 to 4 . . .

- Holman (1973) cites a nucleate boiling correlation by Levy (1959) which indicates that n is 3.

- Rohsenow (1973) states:

  Although the slope is predominantly in the neighborhood of 3 (referring to the slope of q{ΔT} on log log paper and the value of n), observations are available with resulting slopes as low as unity . . . and as high as approximately 25 . . .

- Incropera and Dewitt (1985) point out that an n of 3 is consistent with what is known about nucleation sites.

- Kreith and Bohn (1986) cite a correlation by Rohsenow (1952) which indicates that n is 3, and a correlation by Collier (1981) which indicates that n is 3.33.

WHY LOG LOG PAPER?

Before we appraise the correlation methodology of the old heat transfer, it is appropriate to discuss log log paper. Texts and the literature on the old heat transfer contain a great many graphs, most of which are on log log paper. Why? Why is log log paper the almost exclusive choice for plotting/correlating experimental results in the old heat transfer?

This question certainly seems pertinent, but it is seldom addressed in the old heat transfer. Literature articles do not explain why results are plotted on log log paper, they do not explain why straight lines are drawn through results plotted on log log paper, and they do not explain why it is felt that this methodology leads to valid conclusions about functionality.

Rohsenow (1964) explains why log log paper is used in the old heat transfer:

> *Log log paper is employed . . . to point out the nonlinear relationship even in ranges where it might escape notice on linear paper.*

In other words, Rohsenow's view is that log log paper is used to detect curvature even when there appears to be no curvature on linear paper. This amazing view brings up several questions:

- Doesn't "linear" mean there is no noticeable curvature?
- If there is no noticeable curvature on linear paper, shouldn't you conclude that the relationship is linear?
- If you can't detect the curvature, why worry about it?
- Why plot the data on log log paper if it is a straight line on linear paper?
- If the data is a straight line on linear paper, does plotting it on log log paper somehow affect the functionality inherent in the data?
- Does a straight line on log log paper more rigorously demonstrate power law behavior than a straight line on linear paper demonstrates linear behavior?

Long usage in the old heat transfer makes the unquestioned use of log log paper seem rational. But it is patently obvious to me that, if experimental results closely approximate a straight line on linear paper, we should accept a linear relationship. And we should ignore any conclusion based on plotting the results on log log coordinates or polar coordinates or any other system of coordinates.

LOG LOG PAPER

Every straight line drawn on log log paper passes through the origin if the slope is positive--if the exponent in the corresponding power law is greater than zero. Recall that Eq. (37-1) is the equation of a power law, and note that the power law equation is a particular form of the general equation given by Eq. (37-2):

$$y = mx^n \tag{37-1}$$

$$y = mx^n + b \tag{37-2}$$

Whenever a straight line of positive slope is drawn through data points plotted on log log paper, it is implicitly assumed that b equals zero because Eq. (37-1) is the power law equation for *all* straight lines of positive slope on log log paper. If this assumption is valid, then the power law deduced from the straight line will agree with the functionality of the plotted results (provided there is no bias in the data).

But if this assumption is not valid--ie if the value of b described by the data points is not zero--then the power law deduced from the straight line will *NOT* agree with the functionality inherent in the plotted data. For example, a straight line drawn through data points plotted on log log paper may indicate an exponent of 3 even though the plotted data actually describe an exponent of 1!

In summary, the power law deduced from a straight line of positive slope drawn on log log paper will resemble the functionality described by the plotted data *ONLY* if b is in fact equal to zero. Therefore, in order to demonstrate that the deduced power law does in fact resemble the functionality described by the plotted data, it must somehow be demonstrated that the data indicates b is equal to zero.

WHY NOT LOG LOG PAPER?

There is an impelling reason why nucleate boiling $q\{\Delta T\}$ functionality should not be determined on the basis of straight lines drawn through data plotted on log log paper. A priori evidence demonstrates that boiling requires a *finite* temperature difference--a temperature difference considerably greater than zero. (Recall from a previous chapter that boiling usually requires a wall superheat of 20 or 30 F.) Therefore the boiling function does *NOT* pass through the origin because the function does not even *exist* at the origin!

Since the boiling function does not exist at the origin, and since all straight lines on log log paper pass through the origin if they have positive slope, we conclude there is no good reason to draw straight lines through nucleate boiling data plotted on log log paper. No good reason to suppose that nucleate boiling power law correlations agree with the data they purport to describe. In fact, there is good reason to suppose that nucleate boiling power law correlations do *NOT* agree with the data they purport to describe.

DEFINITIVE RESULTS BASED ON LOG LOG PAPER

Log log paper can be used to obtain definitive results with regard to functionality. However, this requires a special methodology not often used, and not worth the required effort because it is much simpler to obtain definitive results with linear paper rather than log log paper.

In order to obtain definitive results with log log paper, it is necessary to:

- Determine the value of b indicated by the data. This is accomplished by plotting the data on linear paper, drawing a curve through the data, and extrapolating the curve to the y-axis.
- Transform the data so that the additive constant is zero. This is accomplished by subtracting the value of b determined above from the y data.

- Plot the transformed results on log log paper. In other words, plot (y-b) vs x.
- Draw the best straight line through the points.
- Determine the (y-b){x} power law from the straight line.

Using the above procedure, it is possible to use log log paper to generate correlations which agree with the data they purport to describe. But it is not worth the effort. It is much simpler and more reliable to use linear paper to correlate data.

BACKGROUND ON THE NEW WAY

In 1964, I attempted to arrange for the publication of an article which pointed out the vagaries of log log paper and described the manner in which log log paper can be used to obtain definitive results. The article also presented the "new" way to correlate data using linear paper in place of log log paper. The correlation of nucleate boiling data was used for illustration, but only in a generic way.

When the article was reviewed, several reviewers insisted that the article deal with specific nucleate boiling data, so I rewrote the article using data by Berenson (1962) and by Cichelli and Bonilla (1945). The paper was accepted for publication in the A. I. Ch. E Journal on April 21, 1964. Unfortunately, it has not yet been published, even though I received the galley proofs from the Journal 25 years ago!

The following is taken from the galley proof prepared by the Journal preparatory to the planned publication of the article:

> CONCLUSIONS
>
> 1. The nucleate boiling data in the literature do not generally support the contention that the heat flux and the thermal driving force are related in a nonlinear fashion.

2. A reasonably extensive sample of nucleate boiling data taken from the literature suggests that the relationship between heat flux and thermal driving force is essentially linear during nucleate boiling.

3. In general, data should not be plotted on log-log paper unless it is first demonstrated that there is no additive constant such as b in Equation (37-2).

4. Theoretical considerations suggest that nucleate boiling correlations should include an additive constant in order to account for the fact that boiling requires a thermal driving force of finite magnitude.

An abridged version of the manuscript appeared in an advertisement in the April, 1965 issue of Nucleonics. Unabridged copies of the manuscript are available from Ventuno Press without charge.

DATA CORRELATION--NEW WAY

The "new" way to correlate data is :

- Make *NO* a priori assumption about functionality.

- Plot the data on *linear* paper.

- Draw the best curve or straight line through the points.

- Try to determine the simplest expression which closely resembles the curve or straight line through the points. Consider all simple functions including power laws. Even if the curve resembles a power law, do *NOT* plot the points on log log paper. (It is a simple task to draw a power law on linear paper, particularly with computer graphics.)

- Plot the induced correlation on the same linear paper which contains the data points.

NUCLEATE POOL BOILING CORRELATION--"NEW" WAY

When the "new" methodology is used to correlate nucleate boiling data in the literature, it is apparent that the data are highly linear, even though power laws have been used to describe nucleate pool boiling data for 40 years.

It would be desirable to demonstrate that the specific data used to document the power law conclusion is in fact highly linear. Unfortunately, the power law conclusion is usually stated as a generic observation, and no specific data is cited.

In order to demonstrate that nucleate boiling data are quite linear, I wrote the above article using all the nucleate boiling data reported by Berenson (1962). This data is shown in numerous references on log log paper, and with straight lines drawn through the data points. A second reason I used Berenson's data had to do with Fig. (37-1).

Following a suggestion I had made earlier, Rohsenow (1964) plotted Berenson's nucleate boiling data on linear paper, and Fig. (37-1) was his result. He sent me Fig. (37-1) in the course of more or less public correspondence we had with regard to my manuscript mentioned above. In the cover letter which accompanied Fig. (37-1), Rohsenow explained:

> . . . *now I understand how you were led into error on this question of linearity*. . .
>
> *I have enclosed a complete plot of Berenson's data on linear paper and samples of power law curves which are appropriate to the data.*

I do not understand how Rohsenow or anyone else can look at Fig. (37-1) and conclude that the data demonstrate a high degree of curvature.

In my view, Fig. (37-1) speaks loud and clear. What it says is that the data demonstrate linearity. And the points and the curves and the straight lines were *all plotted by Rohsenow or someone following his instructions*.

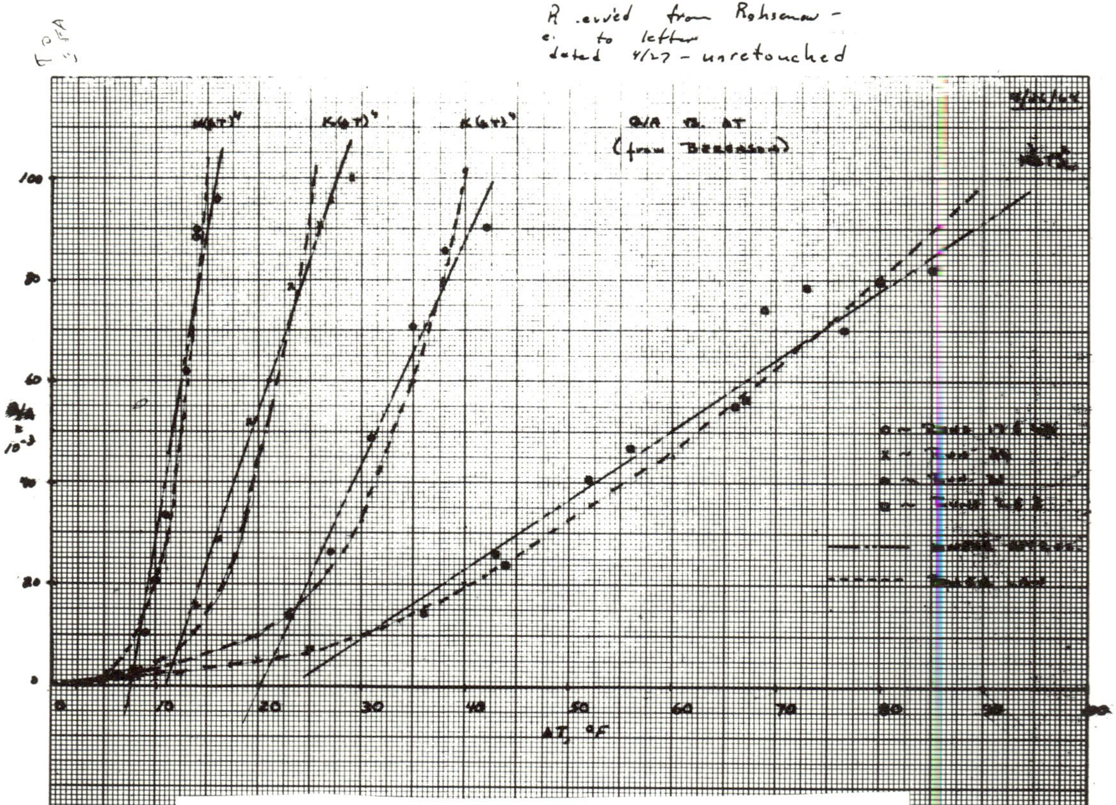

FIG. (37-1) Linear Plot of Berenson's Data--by Rohsenow (1964)

Notice in Fig. (37-1) that there is no curvature evident in the plotted data points. There is a natural tendency to draw the curves through 0,0 because q is zero if ΔT is zero. But in this case, we know that the curves do NOT pass through 0,0 because boiling requires a *finite* ΔT.

If nucleate boiling correlations are intended to describe both the nucleate boiling regime AND the non-boiling regime adjacent to it, then the curves in Fig. 37-1 fill this dual role better than the straight lines.

But if the purpose of nucleate boiling correlations is to describe nucleate boiling behavior rather than non-boiling behavior, then the straight lines are better. And surely the purpose of nucleate boiling correlations has nothing to do with non-boiling behavior.

In short, the data in Fig. (37-1) do *NOT* demonstrate non-linearity, and therefore we should accept that the data in Fig. (37-1) indicate linearity. Since this same nucleate boiling data is widely shown on log log paper with straight lines drawn through the points, and since the slope of these lines suggest power laws with exponents of 3 or 4, it seems quite likely that other nucleate boiling data in the literature also indicate linearity.

THE DATA BY CICHELLI AND BONILLA (1945)

There are a great many nucleate boiling experiments reported in the literature of the old heat transfer, but only a small fraction present q and ΔT data in digital form. One of these is Cichelli and Bonilla (1945).

In the manuscript mentioned above, I plotted nucleate boiling data by Cichelli and Bonilla on linear paper. The plots show no evidence of curvature. When I told Bonilla some time in 1963 or 1964 that his data gave very straight lines when plotted on linear paper, his immediate and definite reply was:

That's impossible!

When I showed him his data plotted on linear paper, he changed his mind.

NEW HEAT TRANSFER METHODOLOGY IN THE OLD HEAT TRANSFER

In 1965, I received a copy of a manuscript which I found exciting and embarassing. Exciting because it concluded that nucleate boiling data are generally linear, and embarassing because it had been published 6 years before the article I had unsuccessfully tried to have published. The manuscript was by Mesler and Banchero (1958).

Mesler and Banchero addressed the problem of functionality in the nucleate boiling regime and stated:

> It has been universal practice to use dimensional analysis to arrive at nucleate-boiling correlations; however, dimensional analysis, as it has been used, usually requires that the data at constant pressure fit (a power law with exponent n). To evaluate n, one plots the data on a log q vs log ΔT plot.

Mesler and Banchero made no a priori judgement about functionality. They simply plotted their data on linear paper. They found that *none* of their nucleate boiling data suggested a ΔT exponent of 3 or 4. They found that *all* their data described straight lines on linear paper. They correctly concluded that the ΔT exponent suggested by the data was essentially *ONE*. And their data included some 600 experimental points obtained at seven different pressures from 15 psia to 515 psia using four different liquids--acetone, ethanol, benzene, and Freon 113.

They also found that the straight lines on linear paper did *NOT* pass through the origin, and therefore any value of n inferred from straight lines on log log paper would have been in error because the data did not describe power laws, since all power laws pass through the origin if n is greater than zero.

Then, in order to determine whether their unusual result was generally applicable, Mesler and Banchero analyzed the data reported by Perry(1940), the data reported by Cichelli and Bonilla (1945), and the data reported by Corty (1951). They found that the data reported by these experimenters also indicated that the ΔT exponent was unity because the data described straight lines when plotted on linear paper.

Carne and Charlesworth (1965) also concluded that nucleate boiling data are generally linear. They state:

> Since the time of the earliest nucleate boiling work it has been customary to present the boiling curves on a log-log basis. . . It is contended here that equally good data presentation is accomplished by plotting nucleate boiling heat transfer results on an arithmetic basis. . . When the data are plotted arithmetically the major portion of the boiling line has the form
>
> $$q = a(\Delta T - \Delta T_0) \qquad (37\text{-}5)$$
>
> where q is the heat flux, ΔT is the wall superheat, . . . and ΔT_0 is the intercept of the boiling line with the abscissa.
>
> Both Berenson and Stock originally presented their data on a log-log basis, but it is evident from (the figures in this report) that plotting the data arithmetically leads to an equally satisfactory, and in the opinion of the authors of this paper a better and more meaningful presentation.

SUMMARY

In summary, the "new" method of correlating data described above is not new in an absolute sense, but only in a relative sense. It is used in isolated instances in the old heat transfer. But it certainly is not the reference method of the old heat transfer which, for some inexplicable reason, largely ignores results obtained using the "new" way and linear paper.

In the new heat transfer, the "new" way is the reference method, and we have no use for the method of the old heat transfer with its assumed power laws and its unshakeable faith in log log paper.

CHAPTER 38

TRANSITION BOILING AND THERMAL STABILITY--PART 1

Transition boiling refers to the region between nucleate boiling and film boiling--the region between the maximum and minimum in the pool boiling curve. In this region, ($dq/d\Delta T$) is *negative*, and this makes transition boiling difficult to deal with because it aggravates the problem of thermal stability.

During the period I was in charge of a boiling liquid metal experimental facility, I discovered that the old heat transfer view of transition boiling and thermal stability is incorrect in several important ways:

- The old heat transfer view is that a boiler which contains a temperature controlled heat source is necessarily stable at all points on the pool boiling curve including the transition region. My work demonstrated that thermal stability is *NOT* assured in this type of equipment, and that thermal instability can cause pronounced hysteresis and undamped oscillatory behavior even if the boiler contains a temperature controlled heat source.

- The old heat transfer view is that boilers in which the power is controlled are necessarily unstable in the transition region. My work indicated that, although power controlled boilers are *generally* unstable in the transition region, they are not *necessarily* unstable in this region. This was an exciting discovery because it indicated that it would theoretically be possible to design power controlled boilers (such as nuclear reactors) which would NOT exhibit a temperature discontinuity at the maximum in the PBC--ie at the "critical heat flux" (CHF) or the "burnout" point.

- The old heat transfer view is that $q\{\Delta T\}$ is highly nonlinear in transition boiling. My work demonstrated that $q\{\Delta T\}$ is highly *linear* in transition boiling. This is important because the magnitude of $dq/d\Delta T$ in transition boiling largely dictates the thermal stability of boiling equipment, and the old view greatly underestimates the minimum value of $dq/d\Delta T$ in the transition region.

This chapter presents a description of transition boiling written 25 years ago and contained in a manuscript which I several times submitted for publication/presentation. The words are as timely today as they were when I wrote them a quarter century ago. Others have since duplicated some of my work on transition boiling and thermal stability, and presented their results in the literature. But heat transfer texts have not yet begun to catch up with my Nucleonics article published in 1964, or with the many manuscripts I submitted for publication and rejection 25 years ago.

The manuscript presented in this chapter is entitled

> Transition Boiling--The Relationship Between
> Heat Flux and Thermal Driving Force

It is dated March 18, 1964, and was first submitted for publication/rejection on March 19, 1964. It discusses the benchmark experiment by Berenson (1962) widely reported and frequently referenced in the texts and literature of the old heat transfer.

Several times, I submitted this article for publication or presentation. It was reviewed and unanimously rejected by pundits of the old heat transfer, including one who had extensive first hand knowledge of Berenson's experiment. His amazing review dated April 15, 1964 stated in part:

> *I believe this man is driving a tack with a sledge hammer. Nothing he says is untrue BUT . . . .*

TRANSITION BOILING--THE RELATIONSHIP BETWEEN HEAT FLUX AND THERMAL DRIVING FORCE

Eugene F. Adiutori

March 18, 1964

ABSTRACT

The relationship between heat flux and thermal driving force has a very strong influence on the stability of boiling equipment in general and on pool boilers in particular. This article examines some of the pool boiling data reported in the literature and concludes that:

> In the transition boiling region of the pool boiling curve, the relationship between heat flux and thermal driving force is highly *linear*.

This examination also indicates that so-called constant temperature boilers are not necessarily stable at all points of the pool boiling curve and that special care must be taken in the design of such boilers in order to enhance their stability.

INTRODUCTION

. . . The specific purpose of this article is to determine the nature of the relationship between heat flux and thermal driving force during transition boiling. This will be done in an entirely empirical manner--ie we will rely only on the experimental evidence quite aside from any theories about the nature of the boiling process. . .

Transition boiling has been investigated by a number of investigators, the most recent being the experiments by Berenson (1962). Because of the excellent precision of Berenson's data and because he controlled so many of the variables which are often uncontrolled, we will deal only with his data. However, because of the depth of his experiments, it seems reasonable to expect the result so obtained to retain a certain degree of generality.

Berenson's data analysis indicated that, during transition boiling, the relationship between heat flux and thermal driving force was highly nonlinear. As described below, this result was obtained (in spite of the fact that the data was highly linear) for two reasons:

- At the time of Berenson's experiment, there was a graphical theory of thermal stability which was incorrect. This theory held that so-called constant temperature boilers could necessarily operate stably throughout the transition boiling region. As a result, Berenson concluded that his data defined the shape of the transition region whereas, in point of fact, he had obtained very little data in the transition boiling region. From the manner in which the data was presented, it would seem that Berenson's equipment could not operate stably in most of the transition boiling region. (A detailed discussion of the stability of Berenson's equipment is beyond the scope of this article. The interested reader is referred to the quantitative theory of thermal stability formulated by Adiutori (reference to Nucleonics article, in publication)).

- Due to the vagaries of log-log graph paper, the lack of data in the transition boiling region was not apparent.

The present analysis of Berenson's data indicates that, in the transition boiling region, the heat flux is linearly related to the thermal driving force.

BACKGROUND

. . . At the time of Berenson's experiment, there was a graphical theory of thermal stability which led to certain conclusions about the stability of pool boilers. With respect to the stability of so-called constant temperature boilers, this theory resulted in the following widely accepted conclusion as stated by Rohsenow (1963):

> With condensing vapor as the heat source on one side of a wall, any point on the entire curve (referring to the pool boiling curve) can be reached under stable conditions.

Since this was precisely the type of boiler used by Berenson, it seemed reasonable to conclude that:

- It was not necessary to design the boiler with stability in mind since it would be inherently stable.

- It was not necessary to verify that sufficient data had actually been obtained in the transition boiling region.

As a result of the second conclusion, the fact that many of the runs had virtually no data in the transition region went unnoticed. The fact that the results were plotted on log-log graph paper also contributed to obscuring the lack of data.

ANALYSIS

Berenson's original analysis of his equipment and experiment are best illustrated by his following statements:

- . . . an experiment designed to control the temperature difference allows operation within the transition region, as well as the other two regions, since there is only one value of heat flux associated with each value of temperature difference.

- Enough datum points were measured in each run to define the characteristic boiling curve completely.

- It was found, with the exception of some of the data presented in Figure 5, that the transition boiling data lie along a straight line connecting the burnout point and the film-boiling minimum point on log-log graph paper. This is also true of the transition boiling data obtained by Braunlich (1941) and Kaulakis and Sherman (1938).

Berenson's first statement agreed with the existing theory of thermal stability but is disproved by the new theory of thermal stability (by Adiutori). (The new theory can be used to prove that three values of heat flux can be associated with each value of temperature difference rather than only one value. The key to this seeming riddle is that the operator can control only the *overall* temperature difference and not the temperature difference corresponding to the thermal driving force between the heat transfer surface and the boiling liquid.)

Berenson's second statement indeed appears to be true when the data is plotted on log-log graph paper (see Berenson's report and also page 119 of Ibele (1963)). However, . . . some of his runs failed to establish the shape in the transition region. For instance, (Runs 17 and 22) contain transition boiling data in the range of heat fluxes between 3,000 and 8,000 Btu/hr ft^2 and no data in the range 8,000 to 90,000. Thus, in these runs, there is no data in over 90% of the transition boiling region! However, (Runs 7 and 9 contained data throughout the transition region and) did indeed establish the complete shape of the pool boiling curve. Thus, Berenson's runs may be classified into two categories:

- Those runs which had essentially no data in the transition region.

- Those runs which contained a more or less continuous set of data throughout the transition region.

We have now to determine which type of run was used to obtain the result that

> The transition boiling heat transfer data plotted as log q vs log ΔT was found to be correlated by a straight line connecting the maximum and minimum heat flux.

Berenson's report contained data for a total of twenty runs. Of these, seventeen could be correlated as Berenson describes above, and the remaining three could not be correlated in the above manner--ie a straight line could not be drawn through three of the runs when the data were plotted on log-log graph paper. In his third statement above, Berenson is referring to these three runs when he states "with the exception of some of the data . . ." Thus Berenson recognized that there were two groups of runs--a large group which could be correlated by straight lines on log log paper and a small group which could not be correlated in this fashion.

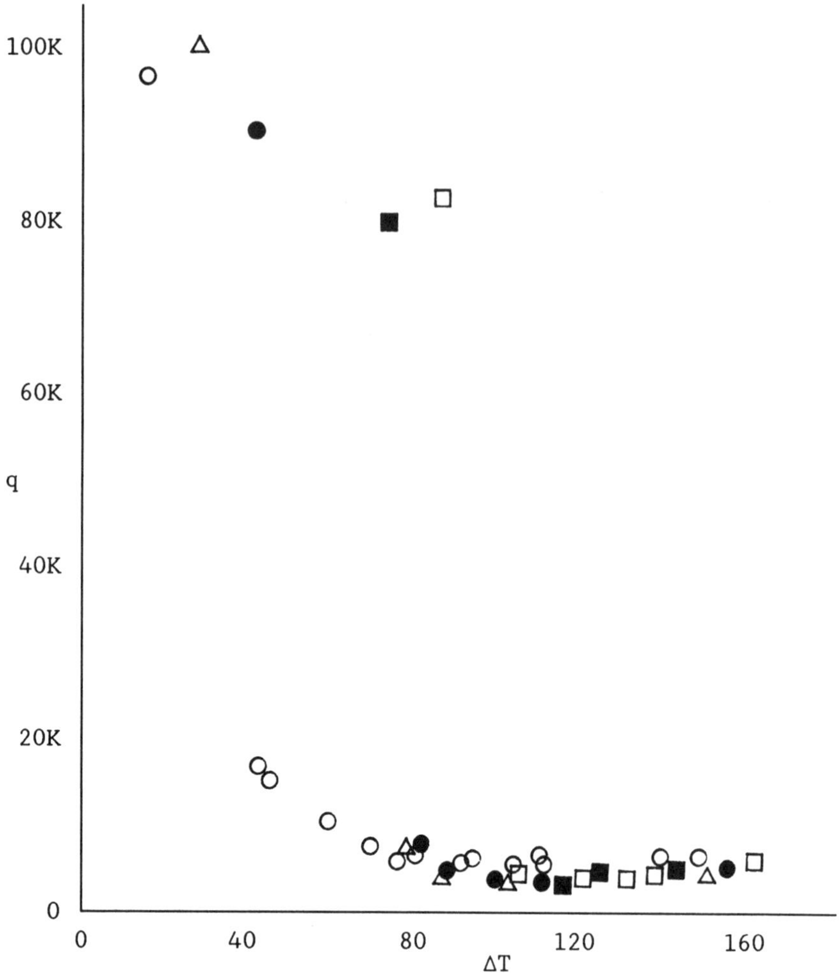

FIGURE (38-1) Transition Boiling Data by Berenson (1962)--Runs 2, 3, 17, 22, 31, 32

(Figure (38-1) is) taken from the large group which seemed to correlate well on log log graph paper. (Figure (38-2) is) taken from the small group which did not correlate. From this, it can be seen that the group that correlated well was the group that contained essentially no data in the transition region! And the small group which did indeed contain the desired data could not be correlated by a straight line on log-log paper! Moreover, it can be seen from (Figures (38-1) and (38-2)) that:

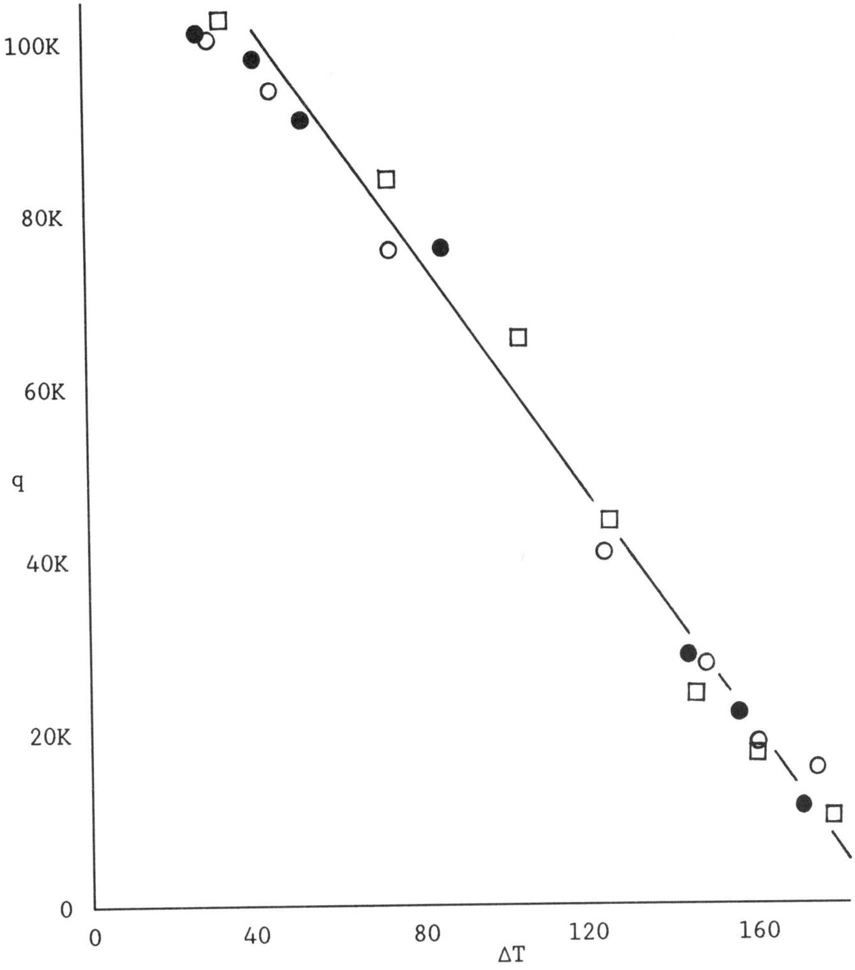

FIGURE (38-2) Transition Boiling Data by Berenson (1962)--Runs 7, 8, 9

- When data was obtained in the transition region, this data was highly linear and would not be expected to exhibit a straight line on log-log graph paper.

- When data was largely not obtained in the transition region, it is of course not possible to state the relationship between heat flux and thermal driving force. However, (Figure (38-1)) seems to suggest that the unmeasured data would also have exhibited a linear relationship between the parameters.

In summary, the above analysis leads to the conclusion that, during transition boiling, the heat flux is linearly related to the thermal driving force. It also indicates that so-called constant temperature boilers are not necessarily stable in the transition region.

DISCUSSION

There are a number of reasons why Berenson's transition boiling data are extremely important. One of these is the fact that his data strongly indicated that the graphical theory of thermal instability was incorrect, since he apparently was *NOT* able to obtain data at all points of the pool boiling curve.

This inability to operate in certain regions was in direct contradiction to the prevalent theory of thermal stability. Moreover, this result of Berenson's serves as virtually indisputable evidence of the rationale of the new theory of thermal stability formulated by Adiutori (reference to Nucleonics article in publication).

(It should be noted that the above states only that Berenson *apparently* was not able to operate in certain regions of the transition region. This qualification is required as a result of the fact that the data published by Berenson is only part of the data he obtained on the boiler.

To state definitively that the absence of data was a result of the inability to operate the equipment in certain regions requires some additional data not included in Berenson's article. This latter point is somewhat aside from the present discussion and will be taken up in a yet unpublished article on hysteresis by the present author.)

Another important result of Berenson's experiment is the realization that a conscious effort must be made in the design stage in order to result in a highly stable boiler. By assuming that any constant temperature boiler would be stable in the transition region, one runs the danger of overlooking the many various design techniques which can be utilized to improve the stability of even a constant temperature boiler.

This point is becoming increasingly important because of the space program and the emphasis being placed on boiling heat transfer to liquid metals. There are several current programs devoted to the construction and operation of pool boilers in the transition region in order to obtain transition boiliing data on liquid metals. Unless these boilers are designed specifically to enhance stability, there is the danger that they also will not be able to operate in this region of transition boiling in spite of the fact that they are constant temperature boilers.

The pronounced linearity evident in (Figure (38-2)) is particularly important from the standpoint of stability. As described by Adiutori (reference to Nucleonics article in publication), the stability of a boiler operating in the transition region is determined in large part by the slope of the heat flux vs. thermal driving force curve. Thus the linearity suggests that all portions of the transition boiling region exhibit approximately the same tendency toward stability or instability.

The result obtained by Berenson that the data could be correlated by a straight line on log-log graph paper would suggest a very large, negative first derivative at the upper end of the transition region. Drawn in this fashion, such a line would overestimate the value of the first derivative by perhaps an order of magnitude. The end result would have been to indicate the virtual impracticability of constructing a boiler which was everywhere stable in the transition region.

A less obvious result of Berenson's experiment is the extreme simplicity of the relationship between heat flux and thermal driving force throughout the nucleate, transition, and film boiling regions. (In order to establish this relationship in a general way), more data should be obtained in the transition region . . .

CONCLUSIONS

The reexamination of Berenson's data supports the following conclusions:

- During transition boiling, the relationship of heat flux to thermal driving force is highly linear.

- Contrary to the graphical theory of thermal stability, so-called constant temperature boilers can not necessarily operate at all points of the pool boiling curve.

- The lack of data in the transition boiling region strongly supports the theory of thermal stability formulated by Adiutori.

- The design of constant temperature boilers intended for operation in the transition boiling region must consider the requirements of thermal stability.

CHAPTER 39

TRANSITION BOILING AND THERMAL STABILITY--PART 2

There was no real understanding of thermal stability until 1964, and consequently the boiler used in Berenson's (1962) benchmark experiment was not effectively designed for operation in the transition region. Until 1964, it was widely held that controlled temperature boilers such as Berenson's could operate stably at any point on the pool boiling curve (PBC), and no design effort was necessary to optimize the thermal stability of this boiler type. This old view is not correct, but it is the view generally presented in current heat transfer texts.

If a boiler is to operate stably in the transition region, a design effort must be made to optimize thermal stability because stability is not assured, even in controlled temperature boilers. Berenson's boiler was usually *NOT* able to operate stably in the transition region, largely because no effort had been made to optimize its design with regard to thermal stability.

Berenson's pool boiler contained a very thick boiler plate (over 2" thick), and this design feature had a negative impact on thermal stability. The end result was that:

- The boiler was thermally unstable in the transition region of most of the PBC's investigated.
- Thermal instability made it impossible to obtain transition region data in most of the PBC's investigated.
- The essential lack of data was unnoticed due to the vagaries of log log paper.
- The results were correlated as though the missing data had been obtained. This led to the conclusion that transition boiling data are highly nonlinear, when in fact the data obtained indicated linearity.

Berenson's experiment provides the background for this chapter which:

- Describes the old heat transfer view of thermal stability generally held at the time of Berenson's (1962) experiment, and still (1989) widely held.
- Points out why the old view of thermal stability is incorrect.
- Describes the heat source data which should be reported in experiments like Berenson's, but is usually omitted.

THERMAL STABILITY--OLD WAY

It is surprising that current heat transfer texts have not generally caught up with the 1960's literature on thermal stability. Heat transfer texts still present the intuitive view of thermal stability which has been widely accepted for 50 years, and which is quantitatively and qualitatively incorrect.

Adiutori (1964a) first assigned the generic name "thermal stability" to the stability of the heat transfer process. Although this name seems a logical choice, it has not yet been generally adopted by the heat transfer community. Heat transfer texts which discuss thermal stability usually avoid calling it by any particular name, and generally bring it up during a discussion of the pool boiling curve, Fig. (39-1). The discussion points out that the heat source in a pool boiler is one of two types:

- Power controlled (also called controlled heat flux or constant power) such as electrical or nuclear.
- Temperature controlled (also called constant temperature) such as a condensing vapor or a flowing fluid.

Power controlled boilers exhibit pronounced hysteresis in the transition region, as shown in Fig. (39-1). Temperature controlled boilers do *NOT* exhibit hysteresis in the transition region. For example, Rohsenow (1963) states:

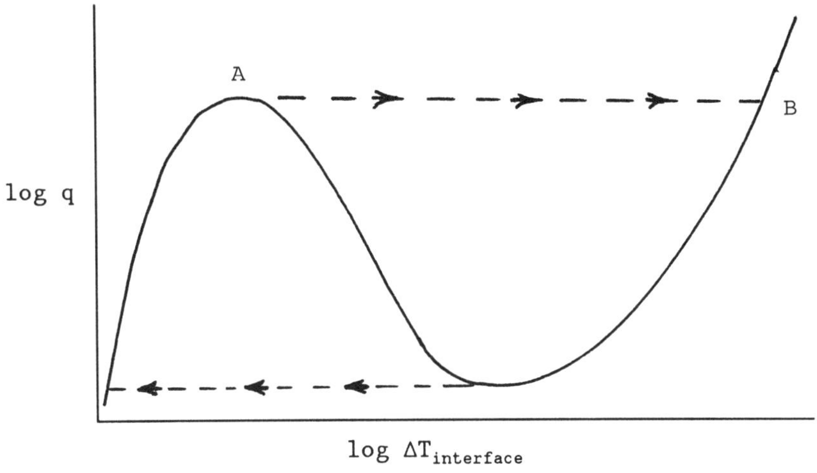

FIGURE (39-1) OLD HEAT TRANSFER VIEW OF POOL BOILING CURVE AND HYSTERESIS IN POWER CONTROLLED POOL BOILERS

With condensing vapor as the heat source on one side of a wall, any point on the entire (pool boiling) curve, Fig. (39-1), can be reached under stable conditions.

Berenson (1962) describes the view of thermal stability which was current in 1962, and which is still the view generally presented in current heat transfer texts:

> The transition region is inherently unstable in experiments in which only the heat flux is controlled owing to the negative slope (of the PBC) in the transition region. However, an experiment designed to control the temperature difference allows operation within the transition region, as well as the other two regions, since there is only one value of heat flux associated with each value of temperature. The most convenient technique for controlling the temperature difference uses a condensing fluid as the heat source.

For example, Kreith and Bohn (1986) discuss the thermal stability of pool boilers. They state:

> If the surface temperature is raised further so that the ΔT at the maximum heat flux (in the PBC) is exceeded,

any of three things can occur, depending on the method of heat control and the material of the heat surface (reference to Westwater (1959)):

1. If the heater surface temperature is the independent variable and the heat flux is controlled by it, the mechanism will change to transition boiling and the heat flux will decrease. This corresponds to operation in (the transition regime).

2. If the heat flux is controlled, as in an electrically heated wire, the surface temperature is dependent on it. Provided the melting point of the heater material is sufficiently high, a transition from nucleate to film boiling will take place, and the heater will operate at a very much higher temperature. This case corresponds to a transition from (Point A to Point B in Fig. (39-1).

3. If the heat flux is independent, but the heater material has a low melting point, burnout occurs. . . .

With regard to thermal stability, Holman (1986) states:

An electrically heated wire is unstable at Point A (in Fig. (39-1)) since a small increase in ΔT at this point results in a decrease in the boiling heat flux. But the wire still must dissipate the same heat flux, or its temperature will rise, resulting in operation farther down the boiling curve. Eventually, equilibrium may be reestablished only at Point B in the film-boiling region. This temperature usually exceeds the melting temperature of the wire, so that burnout results. If the electric-energy input is quickly reduced when the system attains Point A, it may be possible to observe the (transition) region.

(It should be noted that the y axis in Holman's PBC, Fig. 9-3, is labeled log h, but the figure title and the discussion of the figure indicate that it is actually log q. The figure is taken from Farber and Scorah (1948) who presented it in the form log h vs log ΔT. Most authors who include this same figure change the y axis to log q. Also note that the last sentence does not accord with the general old view.)

Incropera and Dewitt (1985) state:

> Nukiyama (1934) was the first to identify different regimes of pool boiling using (an electrically heated wire). The heat flux from a horizontal nichrome wire to the saturated water was determined by measuring the current flow I and potential drop V. The wire temperature was determined from knowledge of the manner in which its electrical resistance varied with temperature. This arrangement is referred to as *power-controlled* heating, wherein the wire temperature . . . is the dependent variable and the power setting . . . is the independent variable. . .
>
> Nukiyama believed that . . . by using a heating process permitting the independent control of ΔT (across the boiling interface), the (transition region missing in his experiment) could be obtained. His conjecture was subsequently confirmed by Drew and Mueller (1937). By condensing steam inside a tube at different pressures, they were able to control the value of ΔT for boiling at the tube outer surface and thereby obtain the missing portion of the boiling curve.

WHAT'S WRONG WITH THE OLD VIEW OF CONTROLLED TEMPERATURE POOL BOILERS?

The old heat transfer view of the thermal stability of controlled temperature pool boilers is based on the view that "the heater surface temperature is the independent variable and the heat flux is controlled by it"--ie that "there is only one value of heat flux associated with each value of temperature". This old view by Kreith and Bohn (1986) and by Berenson (1962) is equivalent to stating that the "heat in" function is a vertical line in Fig. (39-1). If the "heat in" function were a vertical line, this would in fact guarantee thermal stability throughout the transition boiling region.

The difficulty with this old view is that

> *Surface temperature is NEVER the independent variable, and therefore the "heat in" curve in Fig. (39-1) is NEVER a vertical line, even if the heat source is temperature controlled.*

Since the "heat in" curve is not vertical in temperature controlled boilers, this type boiler is *NOT* necessarily stable in the transition region. Thermal instability in this type boiler *OFTEN* causes pronounced hysteresis.

When the heat source is a condensing vapor or a flowing fluid, the temperature of the heat *SOURCE* is often essentially independent. When this is true, the temperature of the boiling *SURFACE* is *dependent* on:

- Source temperature.
- Heat flux.
- Thermal coupling ("thermal resistance" in old heat transfer language) between the heat source and the boiling surface.

In short, the old view of the thermal stability of temperature controlled boilers is not correct because it implicitly assumes infinite thermal coupling between the heat source and the boiling interface. Thermal coupling is *ALWAYS* finite, and that is why the stability behavior of temperature controlled boilers little resembles the old heat transfer view.

THE NEW VIEW OF CONTROLLED TEMPERATURE POOL BOILERS

The "heat in" function describes how the heat flux from the source to the boiling interface responds to the temperature of the interface. In temperature controlled pool boilers, the temperature of the boiling interface depends on heat flux, and therefore the "heat in" function is *NOT* a vertical line in Fig. (39-1).

In new heat transfer language, we determine $q_{in}\{\Delta T_{interface}\}$ in the following way:

- In a manner analogous to the solution of Problem 1 in Chapter 3, analyze the given thermal behavior of the boilerplate and the interface on the heat source side of the plate to determine $q_{in}\{T_{source} - T_{bi}\}$.

- Convert $q_{in}\{T_{source} - T_{bi}\}$ to $q_{in}\{\Delta T_{bi}\}$ by noting that

$$T_{source} - T_{bi} = T_{source} - T_{sat} - \Delta T_{bi} \qquad (39\text{-}1)$$

where "source" refers to heat source, "sat" refers to saturation, and "bi" refers to boiling interface.

(In old heat transfer language, the "heat in" function for a controlled temperature boiler is Eq. (39-2) *IF* h and k are independent of q. If h or k depends on q, Eq. (39-2) is not rigorously correct.)

$$q_{in}\{\Delta T_{bi}\} = (1/h_{source} + L/k)^{-1}(T_{source} - T_{sat} - \Delta T_{bi}) \qquad (39\text{-}2)$$

In order to determine the potential operating points at a given source temperature, draw the "heat in" curve on a figure of the PBC such as Fig. (39-1). If the source side interface and the heated wall exhibit proportional thermal behavior (as is often the case), the simplest way to draw the "heat in" curve is to:

- Note that the difference between source temperature and saturation temperature is the x intercept of the "heat in" function.
- Note that the first derivative of the "heat in" expression with respect to ΔT_{bi} is the slope of the "heat in" function.
- Use the x-intercept and the slope to draw the "heat in" function on Fig. (39-1). The result is Fig. (39-2).

Notice in Fig. (39-2) that:

- The "heat in" line is *NOT* vertical, even though the boiler heat source is the temperature controlled type.
- Boiler heat flux is multi-valued even though the figure describes the behavior of a temperature controlled pool boiler.
- The figure is drawn on *linear* coordinates (as indicated by the 0,0 at the origin). If the figure were on log log coordinates, a linear "heat in" line would be curved.

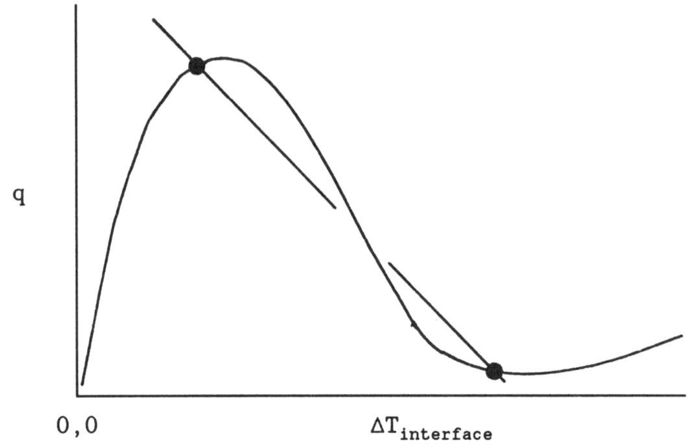

FIGURE (39-2) NEW HEAT TRANSFER GRAPHICAL ANALYSIS OF CONTROLLED TEMPERATURE POOL BOILER

- The "heat in" line in Fig. (39-2) is straight only because we assumed proportional thermal behavior between the heat source and the boiling interface. However, this assumption often applies with reasonable accuracy.

Fig. (39-2) illustrates the new and correct thermal analysis of controlled temperature pool boilers. It shows how to determine whether the heat flux is single-valued or multi-valued at a specific interface temperature, and how to determine the potential steady-state operating points.

Fig. (39-3) contains the same information as (39-2), but it describes the full operating envelope of the boiler rather than a single operating point, and it is in a much more useful form. Notice that Fig. (39-3):

- Describes how the boiler responds to the temperature of the heat source, and this is the parameter of real concern to the designer, the analyst, and the boiler operator.

- Reveals the potential operating points of the boiler over the complete operating envelope *by inspection*.

- Reveals the stability behavior of the boiler over the complete operating envelope *by inspection*.

It is important to recognize that the x axis in Fig. (39-3) is the *OVERALL* temperature difference between the heat source and the saturation temperature of the boiling fluid. This is the temperature difference which is oftentimes truly independent--not the temperature difference across the boiling interface, as in the old heat transfer view.

Fig. (39-3) is readily generated in the following way:

- On Fig. (39-2), draw a set of "heat in" curves which covers the operating envelope of the boiler.
- Note that each "heat in" curve describes q for a fixed value of overall ΔT, and therefore the intersections establish (q, $\Delta T_{overall}$) coordinates.
- Use the coordinates from the preceding step to prepare a q vs $\Delta T_{overall}$ figure such as Fig. (39-3).

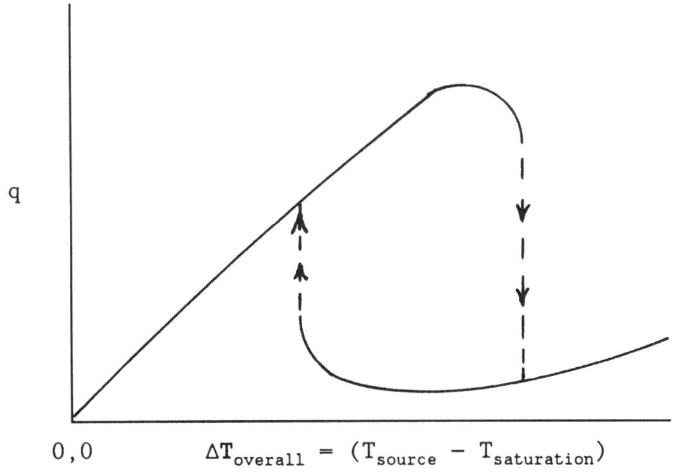

FIGURE (39-3) OVERALL THERMAL BEHAVIOR OF A CONTROLLED TEMPERATURE POOL BOILER WHICH CANNOT OPERATE STABLY AT ALL POINTS ON THE PBC

Fig. (39-3) indicates that even a controlled temperature pool boiler may exhibit hysteresis which prevents operation in much of the transition region. It should also be noted that, if Fig. (39-3) contained no region in which q was multi-valued, this would instantly indicate that the boiler was thermally stable throughout the transition region, and that it would not exhibit hysteresis.

Figure (39-3) describes the "overall thermal behavior" of the boiler. It is the preferred form for describing equipment behavior in the new heat transfer, and was utilized in Adiutori (1965) and in Kovalev (1968).

WHAT'S WRONG WITH THE OLD VIEW OF CONTROLLED HEAT FLUX POOL BOILERS?

The old heat transfer view of controlled heat flux boilers is that q is an independent variable and thus the "heat in" curve in Fig. (39-1) is a horizontal line. The horizontal "heat in" curve indicates that this boiler type is unable to operate stably at any point in the transition region--at any point where $dq_{out}/d\Delta T$ is less than zero.

The old view is 10% true and 90% false. It is true that, if heat flux is the independent variable, the boiler will exhibit the pronounced hysteresis shown in Fig. (39-1). However, this old view is incorrect because:

- *In real boilers, heat flux is seldom an independent variable, even if the heat source is electrical or nuclear.*
- If the boiler plate is resistance heated, the heat flux is independent of the ΔT *only* if the temperature coefficient of electrical resistivity is zero. *It is seldom zero.*
- If the boiler plate is a fuel element in a nuclear reactor, the heat flux is independent of ΔT *only* if the temperature coefficient of reactivity of the fuel is zero. *It is seldom zero.*

To illustrate, suppose an electrical power supply is used to heat a boiler plate by resistance heating, and also suppose that the power supply is a constant current type. In this case, power supply current is the independent variable, and q (ie the heat dissipated in the heater elements) depends on the temperature of the heater elements and their temperature coefficient of resistivity.

If the temperature coefficient is positive, the heat flux increases as the boiler plate temperature increases, and the result is a "heat in" curve with positive slope. The positive slope causes the boiler to become thermally unstable somewhat *BEFORE* the maximum in $q\{\Delta T\}$. The boiler is thermally unstable throughout the transition region, and does not become stable again until somewhat *AFTER* the minimum in $q\{\Delta T\}$. (Recall that the generic criterion for thermal stability is:

$$(dq_{in}/d\Delta T) <s \; (dq_{out}/d\Delta T) \qquad (39\text{-}3)$$

which indicates that thermal stability DEcreases as the value of $(dq_{in}/d\Delta T)$ INcreases.)

The converse of the above is also true--ie if the temperature coefficient of resistivity were negative, stable operation would continue somewhat *beyond* the maximum in $q\{\Delta T\}$, and would return somewhat *before* the minimum.

The same reasoning applies to a fuel element in a nuclear reactor. If the temperature coefficient of reactivity of the fuel is positive, the "heat in" curve has positive slope, and stable operation at the maximum in $q\{\Delta T\}$ is not possible. As before, the converse is also true.

Fig. (39-4) describes the behavior of Nukiyama's (1934) controlled power pool boiler as the power was monotonically increased through the maximum. The figure closely resembles Figure 14 in Nukiyama (1934). Notice that the temperature excursion line is *NOT* horizontal--the heat flux at the end of the excursion was 30% greater than it was at the beginning.

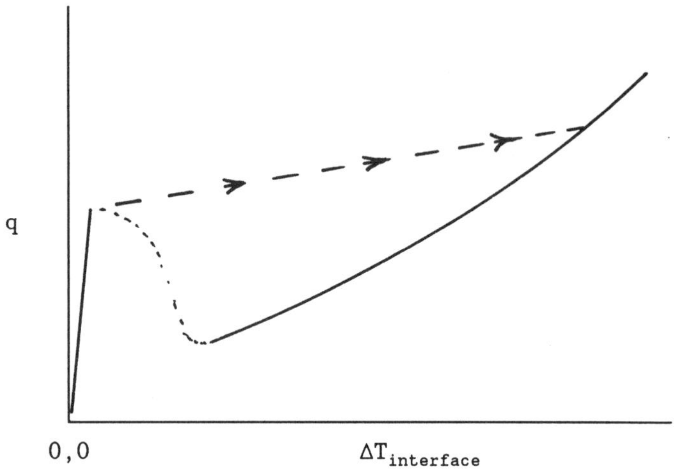

FIGURE (39-4) THE THERMAL STABILITY BEHAVIOR OF THE CONTROLLED HEAT FLUX BOILER USED BY NUKIYAMA (1934)

This line is actually a "heat in" line--it describes how the heat flux into the boiling interface responds to the temperature of the interface. The inclination of this line demonstrates that "heat in" lines are *NOT* necessarily horizontal in power controlled boilers. And since the old heat transfer view of boiler behavior assumes that these lines are horizontal in power controlled boilers, the inclination demonstrates that the old view is incorrect.

The "heat in" lines for Nukiyama's boiler had positive slope, and therefore the boiler became thermally unstable shortly *before* the maximum in $q\{\Delta T\}$, remained unstable throughout the transition region, and did not become stable again until shortly *after* the minimum heat flux.

In summary, the old heat transfer view of the thermal stability of controlled power boilers is simply incorrect. It assumes that, in controlled power boilers, the "heat in" function is independent of the temperature of the boiling interface. This assumption leads to the conclusions that controlled heat flux boilers are unstable in the transition region, and stable throughout the nucleate and film boiling regions. Since this assumption does not accurately describe the behavior of controlled heat flux boilers, the resultant conclusions are both incorrect.

Controlled heat flux boilers are NOT necessarily unstable in the transition region, and they are NOT necessarily stable in the nucleate boiling region.

Moreover, it is within the realm of possibility to design a controlled heat flux boiler which is thermally stable throughout the transition region, and experiences no temperature discontinuity at the maximum in $q\{\Delta T\}$.

THE IMPORTANCE OF HEAT SOURCE/SINK DATA

In heat transfer experiments, it is generally a good idea to obtain and report data which describes the behavior of the heat source and the heat sink. Source/sink data often provides a great deal of insight into the experiment.

With regard to Berenson's (1962) experiment, the primary purpose of the experiment was to experimentally generate pool boiling curves for several fluids and several different surface conditions, with particular emphasis on the transition region. This required the following data:

- Heat flux.

- Saturation temperature of boiling fluid.

- Surface temperature at boiling interface.

Heat source temperature data was not required, but it would have provided valuable insight into the experiment. This data could have been used to generate figures such as Fig. (39-3) which describe the "overall thermal behavior" of the boiler. These figures would have instantly indicated that:

- In the transition region, the boiler heat flux was often a multi-valued function of heat source temperature.

- In the multi-valued regions, the boiler exhibited pronounced hysteresis.

- The hysteresis demonstrated that the boiler was thermally unstable in the region of hysteresis.

- The hysteresis indicated that the complete shape of the subject PBC was not established because hysteresis prevented operation in part of the transition region.

In the old heat transfer, the importance of source/sink data is not generally recognized. For example,the several articles and the doctoral thesis written about Berenson's experiment contain no quantitative or qualitative information about heat source temperature. Yet the experiment was performed at a highly regarded, world class technological institute.

CLOSING REMARKS

It should be noted that this lengthy and seemingly negative discussion of Berenson's experiment is actually a high form of approbation. The experiment was quite comprehensive in several ways, was painstakingly performed, resulted in data with excellent precision, and the data were generously presented in digital form.

It is because Berenson's experiment was excellent in many ways that it is used here to appraise the old heat transfer views and methods. Berenson used the heat transfer science which existed at the time. Could he have done otherwise?

The discerning reader will recognize that my purpose in dwelling on Berenson's experiment is not to find fault with his application of the old heat transfer science.

My purpose is to find fault with the old science.

Berenson's experiment is useful for that purpose, and the fact that the experiment is useful is a tribute to the experimenter.

CHAPTER 40

TRANSITION BOILING THERMAL BEHAVIOR AND THE IMPORTANCE OF FUNCTIONALITY

In the old heat transfer, it is generally assumed that Natural behavior is described by power laws. For that reason, experimental results are generally plotted on log log paper, and correlated by drawing straight lines through the points. This is the method used by Berenson (1962) and other experimenters to determine thermal behavior during transition boiling--ie to determine $q\{\Delta T\}$ in the negative slope region of the pool boiling curve (PBC).

Because Berenson's boiler was not well designed for operation in the transition region, little transition region data was obtained. On the basis of the many runs which contained little or no data in the transition region, it was concluded that $q\{\Delta T\}$ is highly nonlinear in transition boiling. The few runs which contained the desired data indicated linearity, but this was not recognized because the data were plotted on log log paper.

The methodology of the old heat transfer resulted in an experimental conclusion about functionality which was not in agreement with the data. But functionality is not a matter of great concern in the old heat transfer. It is not important to obtain correlations which accurately describe functionality among fundamental parameters. This is evidenced by the fact that literature articles which present experimental results usually demonstrate correlation goodness by comparing the correlation against group parameters.

Unfortunately, this does *NOT* demonstrate that the correlation accurately describes the functionality of each parameter in the group parameters--it merely demonstrates agreement between correlation and groups. It requires a great deal of doubtful assumption in order to reach a conclusion about the functionality of individual parameters based on the demonstrated agreement between correlation and groups.

The new heat transfer recognizes the need to obtain correlations which accurately describe the functionality of each fundamental parameter. Literature articles plot data on linear paper and demonstrate that correlations accurately describe the functionality of each Natural parameter varied in the experiment. Literature articles have little to do with log log plots, and *NOTHING* to do with group parameters such as Nu, Re, f, or h.

In this chapter, the methodology of the new heat transfer is used to determine the functional relationship between q and ΔT in the negative slope region of the PBC. No assumption is made about functionality, and the data are correlated by induction based on linear plots. The importance of functionality is demonstrated by relating several difficulties which arose from the unfounded conclusion that transition boiling data are highly nonlinear.

THE MEANING OF "LINEAR"

In a strict, mathematical sense, "linear" means that the functional relationship between two variables is described by an equation of the form

$$y = mx + b \qquad (40\text{-}1)$$

where the x exponent is precisely 1.000000000. However, our concern is engineering science rather than mathematics. It is *never* possible to empirically demonstrate that two parameters are related in a precisely linear way because the data always includes scatter.

In an engineering sense, it seems reasonable to use the word "linear" when there is no clearcut evidence of nonlinearity. It also seems reasonable to use "linear" even when there is clearcut evidence of nonlinearity, *provided* the magnitude of the nonlinearity is so small that it has no practical importance. In the remainder of this book, "linear" is used in this engineering sense.

TRANSITION BOILING THERMAL BEHAVIOR--NEW WAY

The old heat transfer method of correlating data is to plot the data on log log paper and draw straight lines through the points. This is the method used by Berenson (1962) to correlate the transition boiling data he obtained.

At the time of Berenson's experiment, there was no understanding of thermal stability, and consequently the boiler used in Berenson's experiment was not optimized for stable operation in the transition region. The end result was that little transition boiling data was obtained, and it was concluded that $q\{\Delta T\}$ in transition boiling is highly nonlinear, even though the data indicated linearity.

Hesse's (1973) understanding of thermal stability was adequate to result in a boiler well designed for stable operation in the transition region. This required maximum thermal coupling (ie minimum thermal resistance) between the heat source and the boiling interface. In the test boiler, the boiling surface was the exterior of a thin walled (.03") copper tube, and the heat source was water flowing inside the tube at velocities to 65 ft/sec. (Recall that Berenson's boiler plate was greater than 2 inches thick.) Because the boiler was well designed for stable operation in the transition region, considerable transition region data were obtained and reported.

Good thermal stability design requires a large, negative value of $dq_{in}/d\Delta T$. Hesse reports that his boiler design was able to attain

$$dq_{in}/d\Delta T = -6000 \text{ Btu/hr ft}^2 \qquad (40\text{-}2)$$

This is a factor of 6 better than the value for Berenson's (1962) boiler calculated by Wallis (1965).

(The importance of $dq_{in}/d\Delta T$ was not understood until 1964, and Berenson understandably did not report its value. In the summer of 1965, I was an invited lecturer at a course on two phase heat and mass transfer given at Dartmouth College by Professor Wallis. In my lecture, I derived Stability Criterion (40-3):

$$(1/h_s + L/k) <_s dq_{out}/d\Delta T \qquad (40\text{-}3)$$

where h_s is the "heat transfer coefficient" on the heat source side of the boiler plate, L is boiler plate thickness, and k is boiler plate thermal conductivity. Crit. (40-3) is the thermal stability criterion for a temperature controlled pool boiler (in which the thermal behavior on the source side of the interface is proportional), written in the language of the old heat transfer. Some time after my lecture, Wallis used Crit. (40-3) to calculate the results reported in Wallis (1965).

I also derived Crit. (40-3) at the Open Forum Session, 7th National Heat Transfer Conference, 8/11/64, in a talk entitled "Thermal Stability and Transition Boiling".)

Hesse (1973) reports 11 PBC's obtained with 3 Freon refrigerants at various pressures to 30 atmospheres. The PBC data are shown on six log log plots. In the transition region, the data points indicate pronounced curvature. The shape of the curves in the transition region resembles the shape of Berenson's (1962) few PBC's which were not well correlated by straight lines on log log paper.

(Hesse presents one linear plot. Its purpose is not to describe the PBC data, but rather to explain the impact of thermal stability on the experiment. The figure contains the data from 3 PBC's. In the transition region, the data are quite linear.)

When the new methodology is used to determine the functionality in Hesse's data, the data points describe essentially straight lines on linear paper. The straight lines agree with the new way correlation of Berenson's data. Together, the data by Berenson (1962) and Hesse (1973) indicate that thermal behavior in the transition region of the PBC is generally *LINEAR*.

THE IMPORTANCE OF DIGITAL RESULTS

It should be noted that Hesse's and Berenson's results are particularly useful because they were reported in digital form. It is not possible to read the customary log log plots with reasonable precision, nor should it *EVER* be necessary. More than 40 years ago, McAdams (1948) made the following accurate and important observation:

. . . it is hoped that the authors will publish the original data (ie the data in digital form). . . It is difficult to read these (logarithmic) plotted values with reasonable precision.

Unfortunately, plotting on log log paper is still the base line method for presenting experimental results in today's old heat transfer. For some inexplicable reason, the desirable practice of reporting results in digital form is actively *discouraged!*

The converse is true in the new heat transfer. Literature articles are required to present the data and results in digital form (or to make them readily available by storage in a documentation institute). Linear plots are the base line method for displaying data and results.

THE DESIGN IMPORTANCE OF CORRELATION FUNCTIONALITY

The design of optimum equipment requires correlations which accurately describe functionality among the fundamental parameters of importance. The methodology of the old heat transfer is based on group parameters and log log paper, and this methodology makes it unlikely that correlations will be obtained which accurately describe functionality. This is illustrated by the experimental conclusions that nucleate boiling and transition boiling are highly non-linear, when in fact the data indicated linearity.

In other words, the methodology of the old heat transfer is so ineffective that it did not correctly determine the functionality between just two parameters, even though they were related in a very simple way. It therefore seems highly unlikely that this same methodology could correctly determine the functionality among four or five parameters, particularly if they were related in a nonlinear way.

Boiler design illustrates the importance of functionality. In order to design a boiler to operate stably throughout the transition region of the PBC, the designer must know the minimum value of $dq/d\Delta T$ in the transition region, and he must design the boiler so that $dq_{in}/d\Delta T$ is everywhere more negative than $dq/d\Delta T$ of the PBC.

Kovalev (1968) addresses the problem of designing a pool boiler for stable operation throughout the transition region. He estimates the minimum value of $dq/d\Delta T$ by plotting PBC data on log log coordinates. He states:

> In Fig. 2 are presented the measured values of q vs ΔT (for the PBC). The experiments were not conducted over the temperature range 25 to 80 C since boiling here was unstable. In order to obtain steady-state conditions at any point of the transition region, (Criterion (40-3)) must be satisfied). With transition water boiling at atmospheric pressure the (minimum value of $dq/d\Delta T$ on the PBC) is $\simeq -12000$ Btu/hrft^{2}F. It is hardly (possible) to obtain such great values of the heat transfer coefficient by using water vapour for heating.

Kovalev's Fig. 2 is a log log plot of a PBC. A straight line is drawn through the transition region, including the region 25 to 80 C which contained no data. In the lower part of the figure, the PBC is converted to a plot of $dq/d\Delta T$ vs ΔT. There is a well defined minimum shortly beyond the maximum in $q\{\Delta T\}$--the region which contained no data. Kovalev correctly notes that this minimum establishes the condition at which the boiler is most likely to be thermally unstable.

(Straight lines on log log paper do *NOT* have a constant first derivative as do straight lines on linear paper. If a straight line on log log paper has negative slope, the first derivative is negative and its value increases as x increases--ie the absolute value of the derivative decreases as x increases.)

Kovalev's analysis resulted in the following conclusions:

- The minimum value of $dq/d\Delta T$ in the transition region occurs just beyond the maximum in the PBC.
- Based on the available data for water, the minimum value of $dq/d\Delta T$ in the transition region is estimated to be -12000 Btu/hrft^{2}F.
- It would probably not be possible to design a boiler for stable operation throughout the transition region if steam were the heat source. The reason is that the required heat transfer coefficient could probably not be attained from condensing steam.

It is important to note that:

> *The estimated value of -12000 Btu/hrft^{2}F was greatly influenced by the assumption that transition region data are described by straight lines on log log paper.*

If the estimate had been based on the assumption that transition region data are more or less linear, the same information would have resulted in the following far different conclusions:

- The minimum value of $dq/d\Delta T$ in the transition region does not necessarily occur just beyond the maximum in the PBC. Since $q\{\Delta T\}$ is essentially linear in the transition region, $dq/d\Delta T$ is essentially constant. The end result is that the minimum is not well defined, and may occur any where in the transition region.
- Based on the available data for water, the minimum value of $dq/d\Delta T$ in the transition region may be as large as -3000 Btu/hrft^{2}F. (This value is obtained by assuming a linear change between the given data points at 25 and 80 C, and noting that the lack of data would have resulted if the actual value of $dq/\Delta T$ had been incrementally smaller than -3000 Btu/hrft^{2}F.)
- Since the information about $dq/\Delta T$ is not definitive, it cannot be definitely stated that it is possible to design such a boiler. On the other hand, if $dq/d\Delta T$ is near the upper end of the uncertainty band, it should easily be possible to design such a boiler.

The important point is that the old heat transfer resulted in an unfounded conclusion about functionality in the transition region, and this conclusion caused the designer to grossly overestimate the difficulty involved in designing a boiler to operate stably throughout the transition region. The end result is that, if it were essential to design and build such a boiler, the resultant boiler would be greatly over-designed.

STEPHAN'S (1968) COMMENT ABOUT KOVALEV'S CONCLUSION

Based on the assumption that q{ΔT} in the transition region is described by a straight line on log log paper, Kovalev (1968) concluded that the minimum value of dq/dΔT occurs just beyond the maximum in q{ΔT}, and that thermal instability is most likely at this same point. Kovalev's analysis was correct, but the assumption was incorrect, and therefore the conclusion was also incorrect.

Regarding this conclusion, Stephan (1968) states:

> From his results Kovalev infers that the region of maximum heat flux . . . is the most difficult to investigate. . . This does not occur at the point of maximum heat flux but rather at a point of inflection on the q{ΔT} curve in the region of transition boiling. This was also found in experiments so far carried out by the contributor. May this be confirmed by Fig. 1 which shows the q{ΔT} curve of R114 boiling on the outside of a horizontal tube mounted inside a large vessel.

(Fig. 1 shows a PBC plotted on log log paper. There are many data points shortly beyond the maximum heat flux--the region which Kovalev had concluded should be least stable. The middle of the transition region contains no points, and a dashed line suggests that the boiler was unstable in this region.)

Notice the following in Stephan's comment:

- Stephan offers experimental evidence that the point of minimum thermal stability does *NOT* occur just beyond the maximum in q{ΔT} as Kovelev had concluded, but rather occurs in the middle of the transition region.

- Stephan misinterpreted Kovalev's correct conclusion. Kovalev did *NOT* conclude that "the region of maximum heat flux, ie where the derivative (of the PBC) has a maximum, is the most difficult to investigate". Rather, Kovalev concluded that the point of minimum thermal stability was just beyond the maximum heat flux where the derivative of the PBC was a *minimum*. Kovalev's conclusion is the correct conclusion *if one assumes that the transition region is described by a straight line on log log paper.*

In short, Kovalev's analysis is correct, but the assumption on which it is based is incorrect because it does not agree with transition region data. And because the assumption is incorrect, his analytical conclusion fails to agree with the experimental result obtained by Stephan.

CARNE AND CHARLESWORTH (1965) CONCLUSION ABOUT BERENSON'S TRANSITION REGION DATA

Carne and Charlesworth (1965) also addressed the question of functionality in the transition region. They state:

> There is very little experimental data available in the literature for transition boiling. Berenson (1962) has proposed that transition boiling data can be correlated by a straight line connecting the "burnout" point with the minimum film boiling point on log-log graph paper. With the exception of two runs Berenson's transition boiling data do not extend beyond a heat flux corresponding to 20 per cent of the critical heat flux. On a log log plot the extrapolation recommended by Berenson looks good, with the exception of the two sets of data mentioned above, but the extrapolation does cover 80 per cent of the heat flux range. Figure 10 gives one set of Berenson's transition boiling data, which did comprehensively cover the heat-flux range, and it will be seen that the data correlate excellently rectilinearly on an arithmetic plot.

CHAPTER 41

HOW TO DESIGN EQUIPMENT FOR THERMAL STABILITY

Thermal instability is generally a practical problem only if the equipment includes an interface which behaves in such a way that $dq/d\Delta T$ is less than zero. The common form of this unusual behavior is transition boiling, but it seems likely that the future holds other such forms. (Transition boiling is the negative slope region of the pool boiling curve (PBC) shown in Fig. (41-1).)

There are various ways the designer can deal with potential thermal instability:

- Design the equipment to satisfy the requirements of thermal stability throughout the operating envelope.

- Design the equipment to avoid regions in which it would be thermally unstable.

- Design the equipment so that thermal instability is detected and the equipment shut down before any damage occurs.

- Augment the thermal stability of the equipment by the addition of feedback controllers.

- Augment the thermal stability of the equipment by the addition of an active, parallel heat sink.

- Design the equipment so that it can endure the effects of thermal instability.

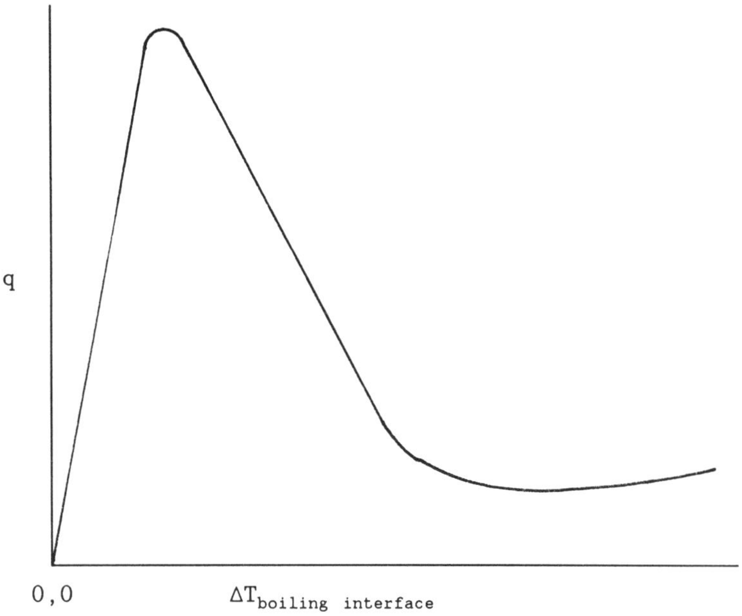

FIGURE (41-1) A Pool Boiling Curve (PBC)

Regions in which $dq/d\Delta T$ is negative may be classified in two groups:

- Regions bounded by endpoints at which $dq/d\Delta T$ is zero.
- Regions bounded by endpoints at which $dq/d\Delta T$ is infinite.

Transition boiling is an example of the first type. The low heat flux region of the nucleate boiling regime for liquid metals is an example of the second type.

(Except for Adiutori (1964a) and the first edition of this book, it is not generally recognized that the undamped oscillatory behavior often observed in pool boiling liquid metals is the result of a negative slope region at the lower heat flux end of the nucleate boiling regime.)

This chapter describes how to design equipment to satisfy the thermal stability criterion throughout the operating envelope. Because the literature contains little reference to negative slope regions bounded by endpoints at which dq/dΔT is infinite, the discussion refers specifically to negative slope regions which resemble the transition boiling region in Fig. (41-1). However, only a minor extension is required to apply the methodology to negative slope regions bounded by endpoints at which the first derivative is infinite. This chapter also summarizes the other methods of dealing with thermal instability listed above, and cites literature examples of these other methods.

Although the discussion often refers to the *boiling* interface, there is nothing in the treatment which is peculiar to boiling. The boiling reference is made because boiling is currently the application of practical interest, and because it simplifies the discussion. However, it should be noted that the discussion is actually quite general.

HOW TO DESIGN EQUIPMENT FOR THERMAL STABILITY

Criterion (41-1) is the generic criterion for thermal stability, and it is the key which makes it possible to design equipment to operate in a thermally stable manner throughout its operating envelope.

$$dq_{in}/dT_{bi} <s \; dq_{out}/dT_{bi} \qquad (41\text{-}1)$$

Criterion (41-1) states that thermal stability is enhanced by small values on the left side, and large values on the right side. In common heat transfer equipment, the left side is negative or approximately zero. Thus thermal instability is a practical problem only when the right side is negative.

If the equipment must operate in a thermally stable manner throughout its operating envelope, Crit. (41-1) must be satisfied at all possible values of T_{bi} (the temperature of the boiling interface). Fortunately, the designer can influence both sides of (41-1).

The right side depends on the boiling fluid and on the condition of the boiling interface--specifically its roughness and cleanliness. Berenson's (1962) results indicated that a rough, dirty surface increases the value of the right side of (41-1) during transition boiling. (Since the right side is negative in transition boiling, an increasing value means a decreasing absolute value.) This experimental result has considerable practical importance because it demonstrates that

> *Quench rate and the rewetting of an uncovered nuclear core can be designer influenced by specifying the roughness of the boiling surface and the trace chemistry of the boiling fluid (in that trace chemistry is an index of "dirtiness").*

The left side of Crit. (41-1) depends only on the design of the equipment on the source side of the boiling interface. The simplest way to establish the value of the left side is to determine an analytical expression for $q_{in}\{T_{bi}\}$, then differentiate the result. The details of the analysis are described in the next few sections.

HOW TO DESIGN FOR THERMAL STABILITY--TEMPERATURE CONTROLLED BOILER

If the boiler has a temperature controlled heat source, dq_{in}/dT_{bi} is determined by analyzing the thermal behavior of the boiler plate and the interface on the heat source side of the plate as shown below:

GIVEN:

The thermal behavior of the interface on the source side of the boiler plate is given by Eq. (41-2). The thermal behavior of the boiler plate is given by Eq. (41-3).

$$q_s = a\,\Delta T \qquad (41\text{-}2)$$

$$q_{bp} = (b/L)\Delta T \qquad (41\text{-}3)$$

ANALYSIS:

$$T_s - T_{bi} = \Delta T_s + \Delta T_{bp} \quad (41\text{-}4)$$

$$T_s - T_{bi} = q_{in}/a + q_{in}L/b \quad (41\text{-}5)$$

$$q_{in}\{T_{bi}\} = (T_s - T_{bi})/(1/a + L/b) \quad (41\text{-}6)$$

$$dq_{in}/dT_{bi} = -1/(1/a + L/b) \quad (41\text{-}7)$$

If heat transfer coefficients had been used to solve the above problem, the given information would have stated that h and k are independent of q, and that h=a, k=b. In place of Eq. (41-7), the result would have been Eq. (41-8).

$$dq_{in}/dT_{bi} = -1/(1/h + L/k) \quad (41\text{-}8)$$

It is important to note that we have obtained the simple result in Eqs. (41-7) and (41-8) because the given information described very simple behavior. In other words, Eqs. (41-7) and (41-8) do not describe temperature controlled boilers in a general way. They describe only those temperature controlled boilers *in which the source side of the boiling interface exhibits proportional thermal behavior.*

For example, if the heat source were a condensing vapor, the thermal behavior of the interface on the source side of the boiler plate would not be proportional. In this case, Eq. (41-8) would be incorrect, and the correct solution would be given by Eq. (41-9):

$$dq_{in}/dT_{bi} = -1/(1/h_{source} + L/k) + \Delta T_{bi}\, d(1/(1/h_{source} + L/k))/dT_{bi} \quad (41\text{-}9)$$

Eq. (41-9) describes controlled temperature boilers in a general way, and it is written in the language of the old heat transfer. If the thermal behavior on the source side of the boiling interface is proportional, the second term on the right is zero, giving Eq. (41-8).

It should be noted that Eq. (41-8) is found in the heat transfer literature, but Eq. (41-9) is the complete form of the equation.

In new heat transfer language, Eq. (41-7) indicates that the thermal stability of the subject boiler is optimized by maximizing the thermal coupling between the heat source and the boiling interface. In old heat transfer language, Eqs. (41-8) and (41-9) indicate that the thermal stability of the subject boiler is optimized by maximizing both h at the source side interface and k/L of the boiler plate.

THE "HEAT IN" CURVE FOR POWER CONTROLLED BOILERS

In the old heat transfer, it is generally stated that power controlled boilers exhibit a temperature discontinuity at the maximum in the PBC. It is explained that the discontinuity results because the "heat in" curve for this type boiler is a horizontal line on PBC figures such as Fig. (41-1).

The difficulty with this old view is that heat flux is not the controlled parameter, even in so-called power controlled boilers.

For example, electrical power supplies may be classified as "constant voltage" or "constant amperage". In either case, if the boiler plate is resistance heated, the power input will depend on the temperature of the boiler plate unless its temperature coefficient of electrical resistivity is zero.

The interaction between electrical resistivity and boiler plate temperature influences the shape of the "heat in" curve. The resultant curve is a horizontal line *only* if the temperature coefficient is zero. (It is seldom zero). It is important to note that:

If the "heat in" curve has positive slope, the temperature discontinuity occurs where the "heat in" curve and the PBC are tangent. This point of tangency occurs BEFORE the maximum in the PBC.

> *If the "heat in" curve has negative slope, the point of tangency and the temperature discontinuity occur AFTER the maximum in the PBC. The discontinuity will not occur at all if the "heat in" curve is everywhere more steeply inclined than the PBC.*

Nukiyama's (1934) Fig. 2 is a generic explanation of the temperature discontinuity which occurs at the maximum in the PBC when heat flux is controlled, and it shows the "heat in" curve as a horizontal line. However, his Fig. 14 is a description of the actual behavior of the test boiler with a platinum wire heating element.

Fig. 14 shows that the actual "heat in" curve had positive slope--the heat flux increased 30% as the wire temperature heated up during an excursion from nucleate boiling to film boiling. During this excursion, the "controlled heat flux" power supply was not adjusted, and the increase was due solely to interaction between the power supply and the temperature of the heating element.

In short, the old heat transfer view is that the "heat in" curve is a horizontal line if the boiler heat source is power controlled, such as electrical or nuclear. The difficulty with this old view is that the "heat in" curve is generally *NOT* a horizontal line in power controlled boilers. This was first recognized by Adiutori (1964a):

> (An) important result of the derivation of the thermal stability criterion is the realization that the designer can vastly improve the stability of equipment by the use of simple design techniques. For instance, it was previously thought that with electrical heat input to the boiler plate . . . it would be impossible to avoid the temperature discontinuity (at the maximum in the PBC). Criterion (41-1) shows that this is not true and that the discontinuity could be avoided altogether by selecting a material with a sufficiently large, positive temperature coefficient of resistivity The required value would be whatever is necessary to satisfy Crit. (41-1) at all points of the pool boiling curve. A number of other equally simple techniques can be used to improve the stability of (nuclear) reactors and other types of heat transfer equipment.

HOW TO DESIGN FOR THERMAL STABILITY--
POWER CONTROLLED BOILER

In order to determine the value of dq_{in}/dT_{bi}, it is necessary to describe how the power input is affected by the temperature of the boiling interface. For example, if the power supply is electrical and the boiler plate is resistance heated, the resistance and the heat input will depend on the temperature of the heating element and ϵ (temperature coefficient of electrical resistivity, 1/F).

If ϵ is positive and the power supply provides a constant voltage, the heat flux will decrease as the temperature of the boiling interface increases, and this will enhance thermal stability. The converse is also true with regard to both the sign of ϵ and the constant voltage or amperage of the power supply.

The quantitative analysis described below is for a constant voltage power supply and a resistance heated boiler plate. It is based on the simplifying assumption that the temperature of the heating element is the same as the temperature of the boiling interface--ie there is infinite thermal coupling between the heating element and the boiling interface.

$$q_{in}\{T_{bi}\} = q_{in}\{T_{sat}\}(1 - \epsilon(T_{bi} - T_{sat})) \qquad (41\text{-}10)$$

$$dq_{in}/dT_{bi} = -\epsilon q_{in}\{T_{sat}\} \qquad (41\text{-}11)$$

$$-\epsilon q_{in}\{T_{sat}\} <_s dq_{out}/dT_{bi} \qquad (41\text{-}12)$$

Crit. (41-12) is the thermal stability criterion for a resistance heated boiler with a constant voltage power supply, based on the simplifying assumption that the temperature of the heating element is the same as the temperature of the boiling interface. Crit. (41-12) states that, for this boiler type:

- Thermal stability in the transition region is possible even though the boiler contains a power controlled heat source.
- Thermal stability in the transition region requires a positive value of ϵ.

- If ϵ is large enough to satisfy Crit. (41-12) throughout the transition region, thermal stability would result throughout the operating envelope.

- If ϵ is negative, the boiler will become thermally unstable *before* the maximum in the PBC (since the right side of Crit. (41-1) equals zero at the maximum).

OTHER WAYS TO DEAL WITH THERMAL STABILITY

There are many ways to deal with thermal stability, and oftentimes it will be desirable to use some method other than designing for stable operation throughout the operating envelope. Other methods include:

- *Design the boiler to avoid the transition region.* In pool boiler design, this is easily accomplished by designing the equipment so that the heat flux cannot reach the value at the maximum in the PBC. This method requires that the application of power be prevented if the boiler does not contain liquid. This is necessary because, if the boiler were dry, it would be easily possible to exceed the temperature at the maximum in the PBC.

- *Design the boiler so that thermal instability is rapidly detected, and the boiler is shut down before any damage occurs.* For example, the temperature of the boiler plate could be monitored and used to shut off the power if the temperature exceeded a predetermined value. The literature contains many examples of this type design.

- *Augment the thermal stability of the boiler by the addition of a feedback controller.* The temperature of the boiler plate is sensed and is used to control the power input to maintain boiler plate temperature. In this way, thermal stability results even though the boiler is not inherently stable. Peterson and Zaalouk (1971) used this boiler type.

- *Install a parallel, inactive heat sink to slow down the temperature excursion which results at the maximum in the PBC.* This gives the operator or the alarm system more time to turn off the power before damage results. Hesse (1973) states that Poletavkin et al (1953) used this boiler type.

- *Augment the thermal stability of the boiler by the addition of an active, parallel sink.* In this boiler design, heat flows to both sinks, even during steady-state operation. Thermal stability is achieved by designing the parallel heat sink so that the total heat flow to the two sinks monotonically increases as the temperature of the boiling interface increases. Ellion (1954) used this boiler type.

- *Design the equipment so that it can endure the effects of thermal instability.* This boiler type is exemplified by Nukiyama's (1934) boiler with the platinum wire in place. Because the platinum wire could endure the temperature excursion which resulted from thermal instability in the vicinity of the maximum in the PBC, he was able to operate in the film boiling region.

  A novel and useful design of this type is the experimental boiler by Harman and Gordy (1967). They state:

  > If the (boiler) heating element had a sufficiently large positive temperature coefficient of resistance it would be thermally self-limiting and data could easily be obtained from both (the nucleate and film boiling) regimes. It is the purpose of this investigation to show that semiconductors have these desirable characteristics and may be advantageously used in the study of boiling heat transfer.

CHAPTER 42

THE NEW POOL BOILING CURVE

The generic pool boiling curve (PBC) is used to promote an understanding of the boiling process. It qualitatively describes the q vs ΔT behavior of a boiling interface in a pool of saturated liquid. The widely accepted PBC in today's old heat transfer accurately describes a maximum and a minimum in $q\{\Delta T\}$, but it fails to describe several other important features of a pool boiling interface.

In the old heat transfer, the generic PBC is defined in a phenomenological way. A curve is presented, and then it is described by the words "pool boiling curve". In the new heat transfer, we take a more rigorous approach. We define "pool boiling curve", and then we determine its shape based on experimental observations reported in the literature.

The literature on the old heat transfer contains a great many articles on pool boiling behavior, but we obtain the required qualitative information from a few comprehensive experiments. The PBC obtained in this new way little resembles the PBC which is widely accepted in today's old heat transfer.

THE DEFINITION OF "BOILING" AND "POOL BOILING CURVE"

"Pool boiling *CURVE*" is seldom defined in today's old heat transfer. Normally, a figure is presented and it is said to be the pool boiling curve. But it would be more rigorous to first define what is meant by "pool boiling curve", and then to determine a pool boiling curve which satisfies this definition and agrees with experimental results reported in the literature.

In preparation for determining the new PBC, let us agree on the following more or less self evident definitions:

"Boiling" is a process in which bubbles are generated in a liquid at or near an interface as the result of heat flow across the interface.

"Pool boiling" is boiling in a pool of liquid through which there is no net flow rate. The pool is usually saturated.

A "pool boiler" is a heated kettle used in pool boiling. It usually includes a reflux condenser which permits extended operation without recharging the kettle.

A "pool boiling curve" (PBC) is the locus of points which relates the heat flux to the temperature difference across the boiling interface in a pool boiler.

It is important to note in these definitions that:

- *If there are no bubbles, there is no boiling.*
- *Where there is no boiling, there is no boiling curve.*

These definitions seem self-evident, but they differ considerably from the old heat transfer view of pool boiling in general and pool boiling curve in particular. For example, the old PBC contains a region in which there are *NO bubbles--NO boiling!*

The old PBC is often presented with the words "boiling curve" located as shown in Fig. (42-1). (See Farber and Scorah (1948), Rohsenow (1963), Kreith (1973), and Holman (1986).) In other words, the old PBC is often presented with the words "boiling curve" written over the PBC in the free convection regime--the very region in which there are *no bubbles* and consequently there is *no boiling!*

Because we have defined "pool boiling curve" to be a description of a *boiling* interface, boiling occurs at all points on the new PBC. Since there is no boiling in the free convection regime of the old PBC, this regime is not part of the new PBC.

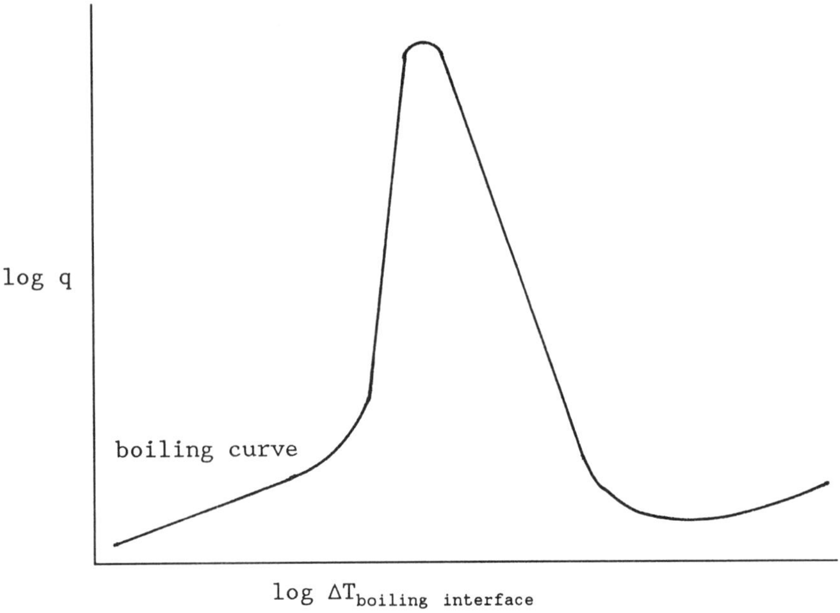

FIGURE (42-1) The Old Pool Boiling Curve

The old pool boiling curve includes boiling and non-boiling modes. It should be referred to as the pool boilER curve, since it describes the behavior of pool boilers, and correctly indicates that they operate in either mode, depending on the heat flux.

MINIMUM SUPERHEAT TO INITIATE OR SUSTAIN BOILING

According to Nukiyama (1984), the 19th century book *Theory of Heat* by Clerk Maxwell and Lord Rayleigh indicates that boiling requires the superheat "at which the cohesion of the water and its contact surface (is) overcome". In other words,

> *Boiling requires a FINITE superheat--ie boiling requires that the temperature difference across the heated interface exceed a minimum value.*
>
> *Data in the literature indicate that this minimum is of the order of 10 to 40 F.*

Nukiyama (1984), in recalling his work with boiling water at atmospheric pressure in the 1930's, states: "I found that the temperature of a metal wire easily reached as high as 105 C without the water boiling." All the pool boiling data presented in Nukiyama (1934) are shown on linear coordinates. Each data set describes two more or less straight line segments connected at a well defined knee, suggesting that the heat transfer mechanism is primarily free convection in the segment below the knee, and well developed nucleate boiling above the knee. In each data set, the knee occurs at $\Delta T \simeq 20$ F.

Corty and Foust (1955) worked with organic liquids. Their Fig. 12 describes the temperature overshoot they repeatedly observed at the transition between the free convection regime and the nucleate boiling regime. They state:

> (If the boiler heat flux was reduced so that) the surface was kept free of bubbles for at least 10 to 15 min., it was possible to retain free-convection heat transfer to much higher ΔT's than would normally be required for vigorous nucleate boiling. Superheats of 40 to 50 F above the saturation temperature of the liquids were possible with no bubbles on the heat transfer surface even though the normal ΔT for violent nucleate boiling was only about 25 F. Such excessive superheats could be maintained for several minutes, but upon further increases the surface spontaneously broke into vigorous nucleate boiling, and the ΔT then decreased to the "normal" value.

Corty and Foust present their boiling data in the form $h\{\Delta T\}$ on linear paper. The same figures also present bubble density data obtained in parallel with the $h\{\Delta T\}$ results. Each data set describes two reasonably straight segments joined at a well defined knee. The figures indicate that:

- The straight segment below the knee is the free convection regime.

- The knee represents the transition between boiling and free convection *as the heat flux is decreased.*

- The region above the knee is well developed nucleate boiling.

- The knee in the $h\{\Delta T\}$ curves coincides with a similar knee in the (bubble density)$\{\Delta T\}$ curves.

- The knees in the h{ΔT} and the (bubble density){ΔT} data occur at a ΔT of approximately 30 F. There are a few bubbles below the knee.

- As the heat flux is decreased below the knee, the last bubble is seen at a ΔT of approximately 20 F. In other words, all boiling ceases at a superheat of 20 F.

The results reported by Schnurmann and Lardge (1973) closely resemble the above results by Corty and Foust (1955). The former state:

> The transition point from convection to nucleate boiling is clearly marked by a sharp drop of the temperature difference between the wire and the bulk of the liquid. . . . At this transition point there may be a temperature drop of nearly 30 C with n-heptane and about 18 C with perfluoromethylcyclohexane.

In summary, the new PBC should indicate that:

- No boiling occurs unless the superheat is finite.

- The inception of boiling often requires a much greater ΔT than steady-state boiling at the same heat flux--ie often involves a temperature discontinuity as the heat flux is monotonically increased.

- The cessation of boiling occurs in a continuous manner. In other words, the surface temperature decreases in a continuous manner as the heat flux is monotonically decreased.

THERMAL BEHAVIOR IN WELL DEVELOPED NUCLEATE POOL BOILING

Thermal behavior in nucleate pool boiling is LINEAR.

Nukiyama (1934) was the first to obtain pool boiling data over a wide range of heat flux. He presents his data on *linear* paper, and each data set describes two reasonably straight line segments connected at a well defined knee. Each data set strongly suggests that the segment above the knee is well developed nucleate boiling.

Many of the segments above the knee are slightly curved. The curvature is generally such that the slope of the line DEcreases as ΔT increases--ie the second derivative is *less* than zero. The curvature is so slight that the curves are straight lines within approximately 1 C.

In the old heat transfer view, the thermal behavior in well developed nucleate boiling is described by a power law with exponent 3 or 4. If this view were correct, the second derivative of $q(\Delta T)$ would be greater than zero, and the slope of the line would INcrease as ΔT increases. Since the slope of Nukiyama's data sets generally DEcreases as ΔT increases, his data describes a second derivative which is *less* than zero, indicating that the ΔT exponent indicated by his data is somewhat less than unity.

Mesler and Banchero (1958) performed a comprehensive pool boiling experiment. They obtained some 600 experimental points at seven pressures from 15 to 515 psia using four different organic liquids. They found that all their nucleate boiling data were well correlated by straight lines on linear paper. The extrapolation of the straight lines intercepted the ΔT axis at positive values. The positive intercepts indicate that boiling requires a finite superheat, and demonstrate that any exponent obtained by plotting nucleate boiling data on log log paper would be in error.

The linear conclusion reached by Mesler and Banchero did not agree with the nonlinear conclusion reached by other experimenters, so they analyzed the data reported by Perry (1940), the data reported by Cichelli and Bonilla (1945), and the data reported by Corty (1951). They found that the data reported by these other experimenters also indicated straight lines when plotted on linear paper.

Adiutori (1964b) analyzed the nucleate boiling data reported by Cichelli and Bonilla (1945) and by Berenson (1962). He states:

> (The analysis of the nucleate boiling data) begins by simply plotting up some experimental results on linear graph paper such as is shown in Figs. 1 to 3. Once the points are plotted and the shape of the curves has become more or less apparent, one must then determine that relationship which seems most strongly suggested by

the data. For the ten runs (by Cichelli and Bonilla (1945) and by Berenson (1962)) shown in Figures 1 through 3, this is a simple matter because the nucleate boiling portions of the curves strongly suggest that straight lines be drawn through the data. Thus, all ten of the runs suggest that the heat flux is essentially linearly related to the thermal driving force by the relationship

$$q = M\Delta T + B \qquad (42\text{-}2)$$

The fact that the several runs obviously exhibit various values for M and B suggests that those constants are not pure constants and that they are indeed functions of such variables as surface condition, pressure, and the many other variables encountered in boiling systems.

It should be emphasized that the scatter which is always present in experimental data makes it impossible to prove that the relationship is precisely linear; that is one cannot infer that the exponent is 1.000000 simply because the data are well correlated by perfectly straight lines. Rather, the data demonstrates that the exponent is not sufficiently different from unity that the difference can be detected. It therefore seems reasonable to accept the value of 1.0 for the exponent, even though any value between 0.9 and 1.1 would probably fit the data equally well.

Close inspection of Figs. 1 to 3 reveals that of the sixty-two nucleate boiling points plotted, virtually every one falls within 2 F of the straight lines drawn through the data. Moreover, the average deviation from the straight lines is less than 1 F. This amazing precision demonstrates that any nonlinearity in the data is of little practical significance.

It is important to note that the data in Figures 1 through 3 are fairly widespread samples of nucleate boiling data. The data consist of data obtained by two different investigators, in two different laboratories, for two different boiler types, for two different boiler plate materials, for three different fluids, for four different pressures, and for five different surface finishes. The fact that every run exhibits a linear relationship between the variables in spite of the widely different nature of the runs strongly suggests that the (linear) results are generally applicable.

> It may perhaps be more convincing to observe that the data in Figs. 1 to 3 would give reasonably straight lines if plotted on log-log paper and that the slopes would indeed be in the neighborhood of 3 or 4. For instance, the same data shown in Figure 3 are also presented by Rohsenow (1963) on page 119 . . . as a log-log graph on which straight lines are drawn through the nucleate boiling data points. On this particular graph, the slope of the lines varies from about 2 to 6, seeming to suggest that the relationship is nonlinear in spite of the fact that it is linear as shown in Figure 3. The reason for this seeming anomaly is presented below. . .

Carne and Charlesworth (1965) analyzed the nucleate boiling data reported by Berenson (1962) for organic liquids and by Stock (1960) for water. These data had originally been plotted on log log paper. Carne and Charlesworth replotted the data on linear paper and noted that the data describe straight lines on linear paper. They state:

> As much, if not more, of the published nucleate boiling heat transfer data can be adequately presented by rectilinear relationships between heat flux and temperature than by log-log plots.
>
> When the data are plotted arithmetically the major portion of the boiling line has the form
>
> $$q = S(\Delta T - \Delta T_0) \qquad (42\text{-}1)$$
>
> where S is the slope of the line and ΔT_0 is the intercept of the boiling line with the abscissa. (They) are constants for a given run but may vary from run to run. . .
>
> Both Berenson and Stock originally presented their data on a log-log basis, but it is evident from (the figures below) that plotting the data arithmetically leads to an equally satisfactory, and in the opinion of the authors of this paper a better and more meaningful presentation.

In short, the data in the literature clearly indicate that the nucleate boiling region of the PBC is *LINEAR* as determined by Nukiyama (1934), by Mesler and Banchero (1958), by Adiutori (1964b), by Carne and Charlesworth (1965), and presumably by others.

THERMAL BEHAVIOR IN TRANSITION POOL BOILING

Thermal behavior in transition pool boiling is generally LINEAR. In other words, q{ΔT} is LINEAR in the region between nucleate boiling and film boiling.

Two benchmark experiments dealing with transition boiling are Berenson (1962) and Hesse (1973). Berenson's experiment was performed before thermal stability was understood. His boiler was not optimized for thermal stability, and he reports little transition region data. The runs which contained little or no transition region data are plotted on log log paper, and it is concluded that q{ΔT} is highly nonlinear in the transition region.

Hesse's (1973) experiment utilized a boiler which was well designed for thermal stability, and he reports a great deal of transition boiling data. The data are plotted on log log paper, and no conclusion is reached with regard to functionality in the transition region.

Hesse's data are also reported in digital form. If the reader will plot Hesse's data on linear paper, he will find that the transition region data indicate a high degree of linearity.

THERMAL BEHAVIOR IN FILM POOL BOILING

Thermal behavior in film pool boiling is the result of radiation and convection from the heated surface which is covered by a blanket of vapor. The radiation is nonlinear, and the convection is more or less linear.

The thermal behavior of film boiling is well behaved, and the above view is the view generally held in the old heat transfer. Holman (1986) and Kreith and Bohn (1986) cite Bromley's (1950) correlation for the convection/conduction component. This correlation indicates that q{ΔT} in film boiling is a power law with a variable exponent which is approximately unity.

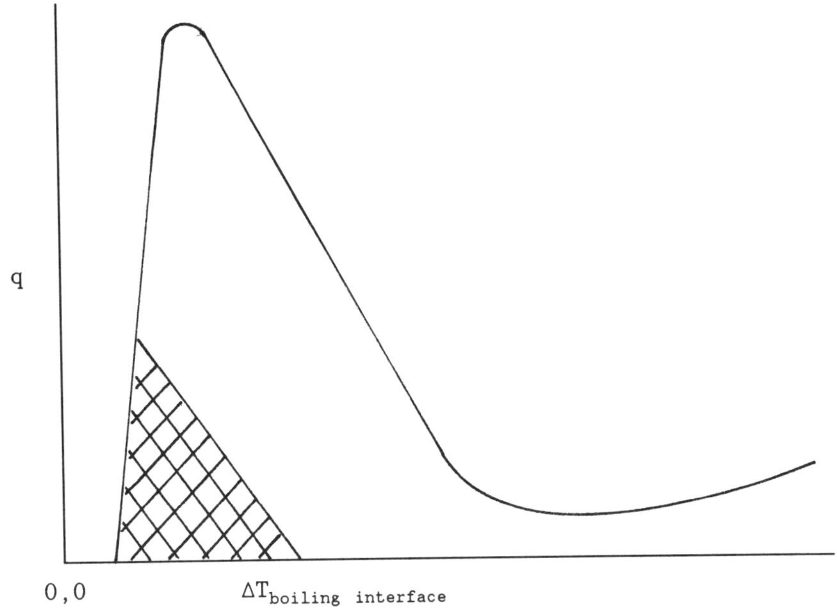

FIGURE (42-2) The New Generic Pool Boiling Curve

THE NEW POOL BOILING CURVE

The new pool boiling curve is shown in Fig. (42-2). Notice the following:

- The new PBC is on *linear* paper.
- There is no region of the new PBC in which boiling does not occur. Boiling requires a *FINITE* superheat.
- The new PBC has a shaded region at the inception of boiling. The shading indicates that the PBC cannot be described in a unique way. The inception of boiling often requires large temperature overshoot, and sometimes results in undamped oscillatory behavior.
- The new PBC is *LINEAR* in the nucleate boiling region.
- The new PBC is *LINEAR* in the transition region.
- The new PBC is unchanged in the film boiling region.

CHAPTER 43

HOW TO CALCULATE THE OVERALL THERMAL BEHAVIOR OF HIGHLY NONLINEAR HEAT TRANSFER EQUIPMENT

The old heat transfer classifies heat transfer equipment on the basis of the heat source. If the heat source is a condensing vapor or a flowing fluid, it is said to be temperature controlled. If the heat source is electrical or nuclear, it is said to be heat flux controlled or power controlled.

There is nothing seriously wrong with these classifications. But the old heat transfer strays far from reality when it concludes that the temperature of a boiling interface is an independent parameter if the boiler heat source is temperature controlled, and when it concludes that the heat flux is an independent parameter if the boiler heat source is power controlled.

The difficulty with the old heat transfer view is that it determines the behavior of the boiler by considering only the nature of the heat source and the thermal behavior of the boiling interface. In the new heat transfer, we recognize that the behavior of real equipment depends not only on these factors, but on other factors as well.

For example, if a pool boiler is heated by a condensing vapor and the temperature of the condensing vapor is independent of heat flux, the old heat transfer view is that the behavior of the boiler is uniquely determined by the PBC. But the new heat transfer view is that the behavior of such a boiler depends on all of the following:

- The thermal behavior of the source side interface (ie the interface on the source side of the heated wall).

- The thermal behavior of the heated wall.

- The shape of the PBC for the fluid and surface condition of interest--ie the thermal behavior of the sink side interface.

In the new heat transfer, we describe the behavior of heat transfer equipment by describing how the heat flux or the heat load responds to the independent parameter in the heat source. We refer to this q{IP} description as the "overall thermal behavior" of the equipment where IP refers to the independent parameter in the heat source.

For example, if the heat transfer equipment is a pool boiler in which the heat source is condensing steam, the temperature of the condensing steam is likely to be an independent parameter. In this case, the "overall thermal behavior" is described graphically by plotting heat flux or heat load vs the temperature of the condensing steam.

If the heat transfer equipment is a pool boiler in which a wall is resistance heated by a constant voltage power supply, the "overall thermal behavior" is described graphically by plotting heat flux or heat load against power supply voltage.

Overall thermal behavior is widely used in the new heat transfer. It is a useful form for describing the behavior of heat transfer equipment because:

- It describes how the heat flux or heat load responds to the parameter which is truly independent and can be set at will by the equipment operator.

- It identifies regions of thermal instability.

- It describes the hysteresis and/or undamped oscillatory behavior caused by thermal instability.

This chapter describes how to use the new heat transfer to determine the overall thermal behavior of equipment. The analysis described here is unnecessary in the old heat transfer because the old heat transfer deals with highly nonlinear heat transfer in a simplistic way which leads to results which are incorrect.

OVERALL THERMAL BEHAVIOR--OLD WAY VS NEW WAY

The "overall thermal behavior" of heat transfer equipment refers to the manner in which the heat flux or heat load responds to the independent parameter in the heat source as it is monotonically increased and decreased through the entire operating envelope.

The old heat transfer incorrectly assumes that the nature of the heat source and the thermal behavior of the sink side interface uniquely determine overall thermal behavior. For example, texts on the old heat transfer use the pool boiling curve to explain that electrically heated boilers experience a temperature discontinuity at the maximum in the PBC, and that boilers heated with condensing steam experience no such discontinuity.

In the new heat transfer, we recognize that this old view is not correct. Electrically heated pool boilers do not necessarily experience a discontinuity at the maximum in the PBC, and boilers heated with condensing steam can and oftentimes do experience a similar discontinuity which occurs beyond the maximum in the PBC.

There is an important design point here which must be noted. It is that, in the old heat transfer view, the designer cannot design against the discontinuity if the heat source is electrical or nuclear because the discontinuity is a necessary consequence of this heat source type. The old view also holds that the designer need not design against the discontinuity if the heat source is condensing steam or flowing fluid because continuity is assured with this heat source type.

In the new heat transfer, we recognize that the temperature discontinuity is *NOT* a necessary consequence of an electrical or nuclear heat source. The temperature discontinuity does *NOT* necessarily occur at the maximum in the PBC. It may occur before the maximum or after the maximum *or not at all*. The designer *CAN* moderate or perhaps even eliminate the discontinuity, even though the heat source is electrical or nuclear.

We also recognize that the temperature discontinuity can occur even if the heat source is condensing steam or flowing fluid, and therefore the designer must always design against the discontinuity in the likely event that the discontinuity is not acceptable.

In the old heat transfer, no design effort is made to eliminate the temperature discontinuity because it is judged to be an impossible or unnecessary task, depending on the heat source type. In the new heat transfer, a design effort is often made to eliminate the discontinuity because we recognize it is *NOT* an impossible or unnecessary task. The important ramifications of this improved understanding are self evident.

OVERALL THERMAL BEHAVIOR--TEMPERATURE CONTROLLED HEAT SOURCE

If a pool boiler contains a condensing vapor heat source, and if the saturation temperature of the heat source is essentially independent as is usually (but not always) the case, its overall thermal behavior is described by the function

$$q\{T_s - T_{sat}\}$$

where the subscript "s" refers to the heat source. In the new heat transfer, this function is obtained from the following thermal behavior functions:

- $q\{\Delta T_s\}$--thermal behavior of the source side interface.
- $q\{\Delta T_{bp}\}$--thermal behavior of the boiler plate.
- $q\{\Delta T_{bi}\}$--thermal behavior of the boiling interface.

The overall thermal behavior is determined from this information by noting that:

$$\Delta T_t = \Sigma \Delta T_i \qquad (43\text{-}1)$$

$$(T_s - T_{sat}) = \Delta T_s + \Delta T_{bp} + \Delta T_{bi} \qquad (43\text{-}2)$$

If all three q functions listed above are given analytically, they are substituted for the ΔT functions on the right side of Eq. (43-2). The resultant equation is solved for q, just as in the Chapter 3 problems. The q function which results is a description of overall thermal behavior, since it is in the desired form $q\{T_s - T_{sat}\}$.

If one of the three q functions is so nonlinear that it is described graphically, Eq. (43-2) is solved as illustrated in Problem (43-1) below.

If the q function for the boiling interface includes a region in which $dq/d\Delta T$ is less than zero, it will be necessary to analyze for thermal instability. Recall that thermal instability can result in either hysteresis or undamped oscillatory behavior.

Hysteresis is readily detected as follows:

- Plot $q\{T_s - T_{sat}\}$ on linear paper.
- If q is everywhere a single valued function, there is no hysteresis.
- Wherever q is a triple valued function, hysteresis results in that region.

Undamped oscillatory behavior cannot be detected on the basis of a multiple-valued $q\{T_s - T_{sat}\}$. It must be detected by applying the criterion for thermal stability--by comparing $dq_{in}/d\Delta T$ against the slope of the PBC. Where the thermal stability criterion is not satisfied *and there are singular intersections* between the "in" and "out" q functions, undamped oscillatory behavior will result. (Intersections which are both unstable and singular can result *only* if the negative slope region is bounded by endpoints at which $dq/d\Delta T$ is infinite.) It is identified on the figure of overall thermal behavior by an oscillating line with q limits which correspond to the q values at the endpoints of the negative slope region.

Problem (43-1) illustrates the determination of the overall thermal behavior of a heat exchanger which contains a highly nonlinear interface on the sink side of the heated wall. Notice that there is nothing in the problem which is peculiar to boiling, and that the problem deals with forced convection equipment.

However, because the temperature change in the fluids is small and can be neglected, the overall thermal behavior of the heat exchanger is exactly the same as that of a pool boiler. Notice that the problem would be unchanged if Fig. (43-1) showed a pool boiler where Fluid A was condensing steam and Fluid B was boiling.

The problem is first stated within the framework of the old heat transfer--ie in terms of heat transfer coefficients. The problem is then stated and solved within the framework of the new heat transfer.

> *The reader is challenged to solve Problem (43-1) using the old heat transfer.*

PROBLEM (43-1) THE OVERALL THERMAL BEHAVIOR OF A HEAT EXCHANGER WITH A CONTROLLED TEMPERATURE HEAT SOURCE--OLD HEAT TRANSFER

Describe the overall thermal behavior of the double pipe heat exchanger in Fig. (43-1). In other words, describe how the total heat transferred is related to the inlet temperature of Fluid A over the complete operating envelope of the equipment. Indicate any hysteresis and any regions of undamped oscillatory behavior.

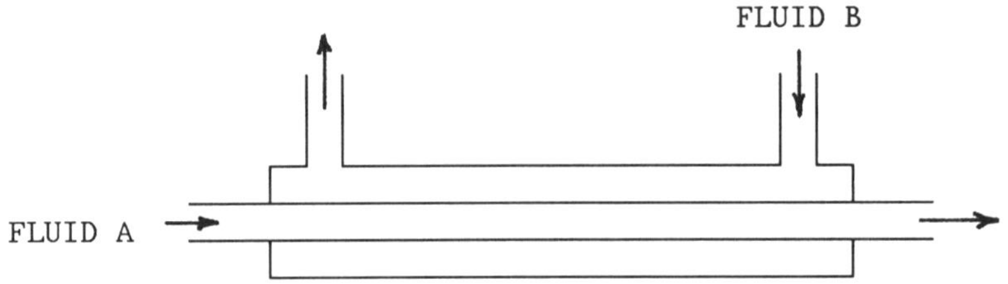

FIGURE (43-1) HEAT EXCHANGER FOR PROBLEM (43-1)

<u>GIVEN</u>

- The heat transfer coefficient for Fluid A is 10000 Btu/hr ft^2 F. The inlet temperature of Fluid A is set by the equipment operator. The operating range for Fluid A is 100 to 250 F.
- The heat transfer coefficient for Fluid B is described in Fig. (43-2). The Fluid B inlet temperature is 100 F.
- The temperature change in Fluids A and B is small and may be neglected.
- The i.d. of the inner pipe is 1.00". Its o.d. is 1.20".
- The thermal conductivity of the inner pipe wall is 9.57 Btu/hr ft F.
- The active area of the heat exchanger is 4.2 ft^2 based on the outer surface of the inner pipe.

The solution of this problem using today's old heat transfer and heat transfer coefficients is left to the reader.

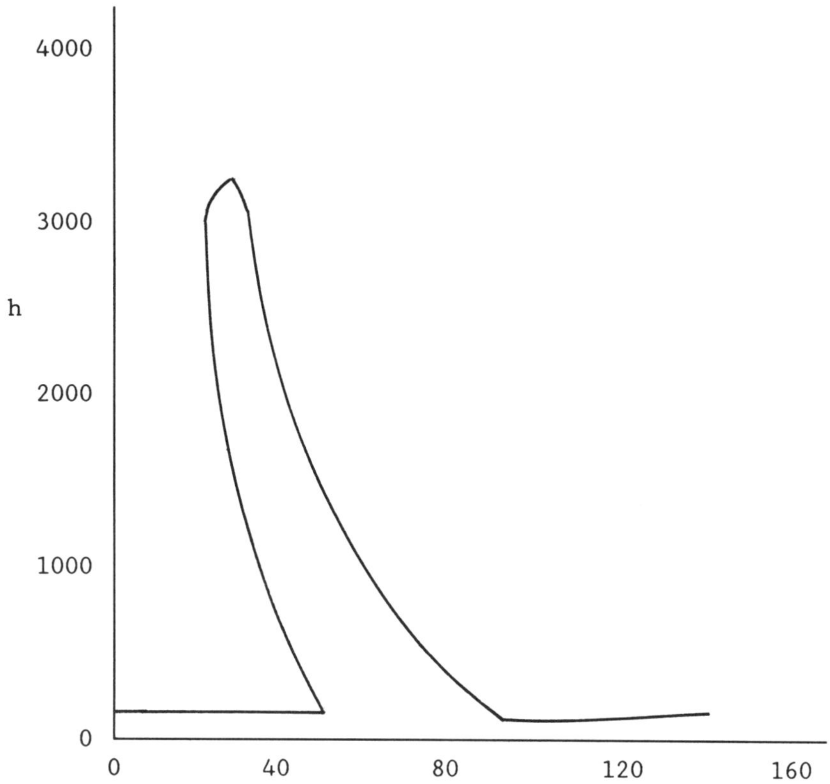

ΔT_B

FIGURE (43-2) h{ΔT_B}, Fluid B, Problem (43-1)

ΔT_B, h COORDINATES FOR FIGURE (43-2)

0, 129	25, 3240
20, 129	30, 3033
40, 129	40 2075
48, 129	60, 883
45, 267	80, 313
40, 525	90. 111
35, 857	100, 124
30, 1333	120, 139
25, 2000	140, 161
20, 3000	

PROBLEM (43-1) THE OVERALL THERMAL BEHAVIOR OF A HEAT EXCHANGER WITH A CONTROLLED TEMPERATURE HEAT SOURCE--NEW HEAT TRANSFER

Describe the overall thermal behavior of the double pipe heat exchanger in Fig. (43-1). In other words, describe how the total heat transferred is related to the inlet temperature of Fluid A over the complete operating envelope of the equipment. Indicate any hysteresis and any regions of undamped oscillatory behavior.

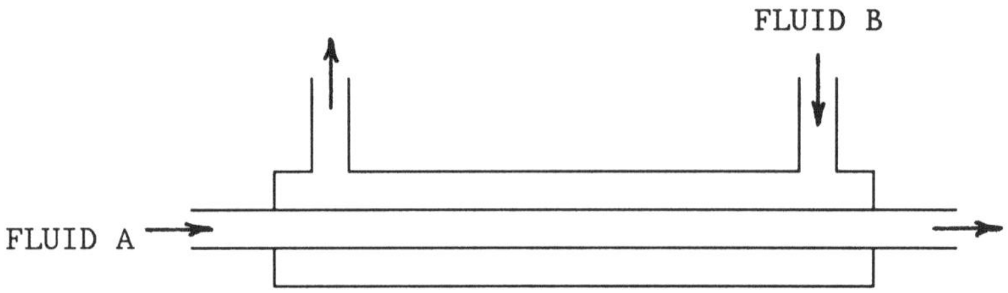

FIGURE (43-1) HEAT EXCHANGER FOR PROBLEM (43-1)

GIVEN

- The thermal behavior of the Fluid A interface is described by $q = 10000\ \Delta T$. The inlet temperature of Fluid A is set by the equipment operator. The operating range for Fluid A is 100 to 250 F.

- The thermal behavior of the Fluid B interface is described by Fig. (43-3). The Fluid B inlet temperature is 100 F.

- The temperature change in Fluids A and B is small and may be neglected.

- The i.d. of the inner pipe is 1.00". Its o.d. is 1.20".

- The thermal behavior of the inner pipe material is described by $q = 9.57\ \Delta T/L$.

- The active area of the heat exchanger is 4.2 ft^2 based on the outer surface of the inner pipe.

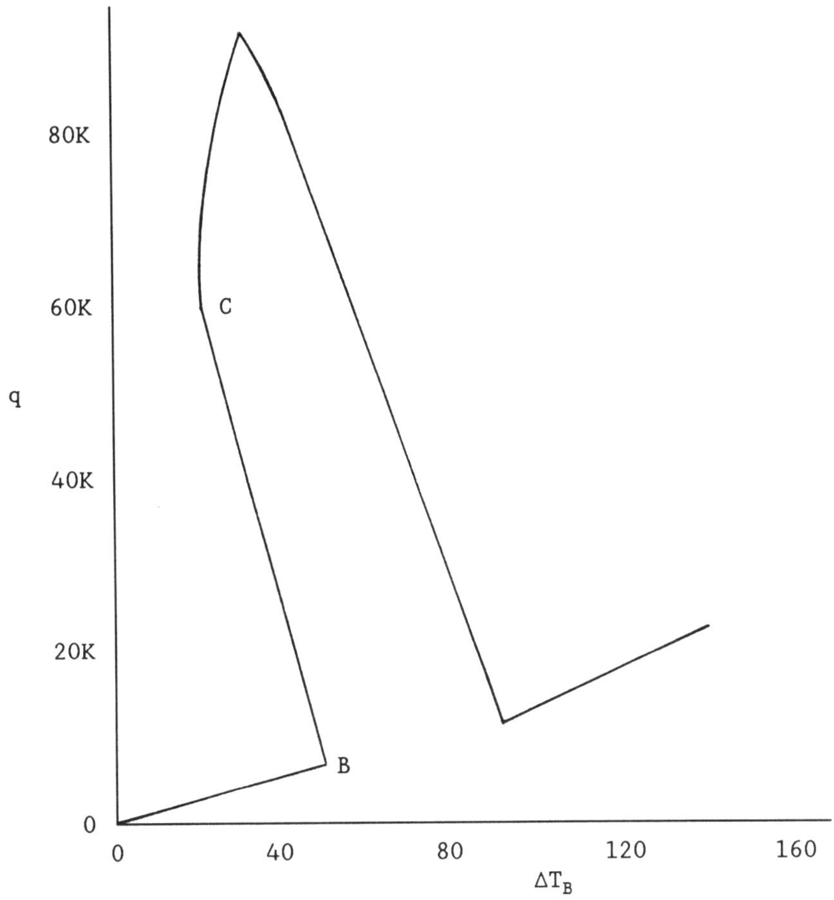

FIGURE (43-3) $q\{\Delta T_B\}$, Fluid B, Problem (43-1)

$\Delta T_B, q$ COORDINATES FOR FIGURE (43-3)

0, 0	25, 81000
20, 2580	30, 91000
40, 5160	40, 83000
48, 6200	60, 53000
45, 12000	80, 25000
40, 21000	90, 11000
35, 30000	100, 12400
30, 40000	120, 16700
25, 50000	140, 22500
20, 60000	

ANALYSIS OF PROBLEM (43-1)

Using the new heat transfer, we note that:

$$\Delta T_t = \Delta T_A + \Delta T_w + \Delta T_B \qquad (43\text{-}3)$$

$$T_A = T_B + \Delta T_t \qquad (43\text{-}4)$$

$$\Delta T_t\{q_o\} = \Delta T_A\{q_o\} + \Delta T_w\{q_o\} + \Delta T_B\{q_o\} \qquad (43\text{-}5)$$

$$\Delta T_A\{q_o\} = q_o d_o/10000 d_i = .000120\ q_o \qquad (43\text{-}6)$$

$$\Delta T_w\{q_o\} = q_o(d_o/2)(\ln d_o/d_i)/(9.57\cdot 12) = .000953 q_o \qquad (43\text{-}7)$$

$\Delta T_B\{q_o\}$ = as described in Fig. (43-3)

where d refers to the diameter of the inner pipe, the subscripts "i" and "o" refer to the inner and outer surfaces of the inner pipe, and "t" refers to total.

Eq. (43-5) contains the function we wish to determine. Since the behavior of the Fluid B interface is described graphically, the simplest way to solve Eq. (43-5) and to determine the overall thermal behavior is:

- Select a set of $(q, \Delta T)$ values along the graphical function, Fig. (43-3).
- At each of the set of q values, use Eqs. (43-6) and (43-7) to solve for ΔT_A and ΔT_w.
- Use the results to plot $q\{T_A\}$, the overall thermal behavior.

The details of the analysis are given in Table (43-1).

The coordinates in Table (43-1) completely describe the overall thermal behavior of the heat exchanger in Fig. (43-1) *provided* the thermal behavior of the nonlinear interface contains no negative slope region bounded by endpoints at which $dq/d\Delta T$ is infinite. If there is such

TABLE (43-1)

DETERMINATION OF COORDINATES FOR FIG. (43-4)

q_o	ΔT_A	ΔT_w	ΔT_B	ΔT_t	T_A	Q
0	0	0	0	0	100	0
6200	1	6	48	55	155	26K
20000	2	19	40	61	161	84K
40000	5	38	30	73	173	168K
60000	7	57	20	84	184	252K
80000	10	76	25	111	211	336K
91000	11	87	30	127	227	382K
80000	10	76	42	128	228	336K
60000	7	57	55	119	219	252K
40000	5	38	69	112	212	168K
20000	2	19	84	105	205	84K
11000	1	10	90	101	201	46K
16000	2	15	117	134	234	67K
22000	3	21	138	162	262	92K

- ΔT_A is from Eq. (43-6)
- ΔT_w " Eq. (43-7)
- ΔT_B " Fig. (43-2)
- ΔT_t " Eq. (43-3)
- T_A " Eq. (43-4)
- $Q = 4.2\ q_o$

a region, the analysis in Table (43-1) does not determine whether thermal instability in this region will result in undamped oscillatory operation, except by default.

The default method of determining undamped oscillatory behavior depends on the following:

> Negative slope regions bounded by endpoints at which $dq/d\Delta T$ is infinite *necessarily* result in either hysteresis or undamped oscillatory behavior.

Therefore, if such a region does not exhibit hysteresis, it must exhibit undamped oscillatory behavior. With regard to the results in Table (43-1), if a plot of q vs T_A does not indicate hysteresis in the BC region, then the BC region exhibits undamped oscillatory behavior. During the oscillations, the heat flux varies between its values at the endpoints of the negative slope regions.

A direct way to detect regions of undamped oscillatory behavior is to calculate the value of the first derivative of the "heat in" function, and then compare the result with the value of the first derivative in the negative slope regions bounded by endpoints at which the derivative is infinite. If the derivative of the "heat in" function is greater than that of the negative slope region (ie if the "heat in" function is less inclined than the "heat out" function), then the region exhibits undamped oscillatory behavior.

In Problem (43-1), the derivative of the "heat in" function is obtained as follows using the new heat transfer:

$$T_A - T_{w,o} = \Delta T_A + \Delta T_w \qquad (43\text{-}8)$$

$$T_A - T_{w,o} = (.000120 + .000953)q_{o,in} \qquad (43\text{-}9)$$

$$dq_{o,in}/dT_{w,o} = -932 \qquad (43\text{-}10)$$

Fig. (43-3) indicates that, in the BC region, the first derivative is approximately -2000 Btu/hr ft^2 F, and therefore this region is thermally unstable. Inspection of Fig. (43-3) indicates that the intersections of the "heat in" and "heat out" functions are single-valued throughout the BC region. Therefore the equipment exhibits undamped oscillatory behavior in the BC region as it endlessly hunts for a stable operating point it is unable to find.

The time dependent behavior of the heat exchanger when operating in the BC region is described in Chapter 36.

ANSWER TO PROBLEM (43-1)

Fig. (43-4) is the new heat transfer answer to Problem (43-1). It is obtained in the following way:

- Plot the (Q,T_A) coordinates in Table (43-1).
- Indicate that the heat exchanger operates in an undamped oscillatory manner in region BC of Fig. (43-3) by showing the heat load in this region as an oscillating line which varies between the Q values at B and C.
- Indicate hysteresis in regions where Q is multi-valued using conventional hysteresis notation.

DISCUSSION OF PROBLEM (43-1)

Fig. (43-4) describes the overall thermal behavior of a one phase, forced convection heat exchanger. But because the bulk temperature change in the fluids is small, Fig. (43-4) also describes the behavior of a pool boiler with a condensing vapor heat source.

Notice that the behavior described by Fig. (43-4) bears no resemblance to the old heat transfer view that

> *With condensing vapor as the heat source on one side of a wall, any point on the entire PBC can be reached under stable conditions.*

Fig. (43-4) demonstrates that, even with condensing vapor as the heat source on one side of a wall, thermal instability can cause pronounced hysteresis and/or undamped oscillatory behavior.

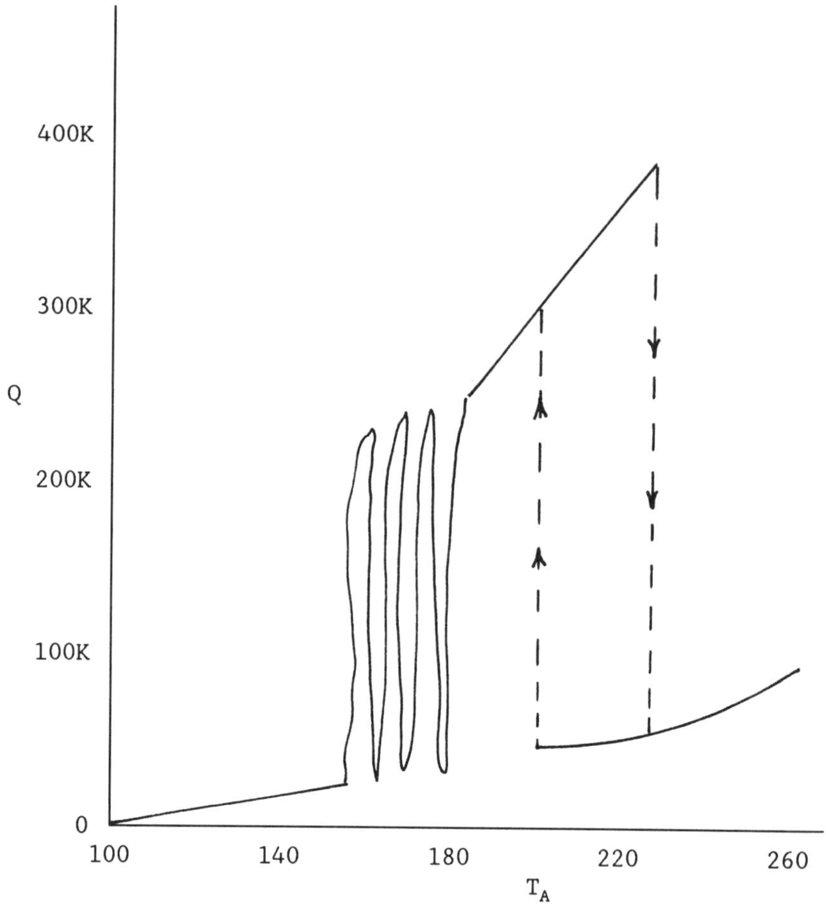

FIGURE (43-4) Overall Thermal Behavior of Boiler in Problem (43-1),

The old heat transfer view would be correct only if the overall thermal behavior were identical to the PBC--only if the thermal coupling between the heat source and the boiling interface were infinite--only if the source side heat transfer coefficient and the thermal conductivity of the heated wall were infinite. And since the thermal coupling between the heat source and the boiling interface is *NEVER* infinite, the old view *NEVER* accurately describes the behavior of real boilers.

Neither Berenson (1960) nor Berenson (1962) presents any data which describe how the heat flux or heat load respond to the temperature of the heat source. (From the manner in which the data are presented in the doctoral thesis, it seems likely that heat source temperature was not included in the test data.) Hesse (1973) also presents no data on heat source temperature. He appraises the thermal stability of the boiler by comparing slopes in the transition boiling region with calculated values of $dq_{in}/d\Delta T$.

In the new heat transfer, we recognize that the simplest and most concrete way to appraise thermal stability in the transition boiling region is to demonstrate that the overall thermal behavior is a single valued function. This requires merely that the experiment be designed so that the test matrix includes a series of steady-state test points during a monotonic increase and decrease in heat source temperature, and that heat source temperature be included in the test data.

In summary, Problem (43-1) demonstrates how to determine the overall thermal behavior of heat transfer equipment which contains a controlled temperature heat source. Overall thermal behavior is an important part of the new heat transfer because it accurately and succinctly describes the behavior of highly nonlinear equipment. It is not part of the old heat transfer because the old heat transfer incorrectly determines boiler behavior by considering only the nature of the heat source and the PBC.

PROBLEM (43-2) THE OVERALL THERMAL BEHAVIOR OF A HEAT EXCHANGER WITH CONTROLLED POWER HEAT SOURCE

Describe the overall thermal behavior of the heat exchanger in Fig. (43-5). In other words, describe how the heat flux is related to the power supply over the entire operating envelope of the heat exchanger. Indicate any hysteresis and any regions of undamped oscillatory behavior.

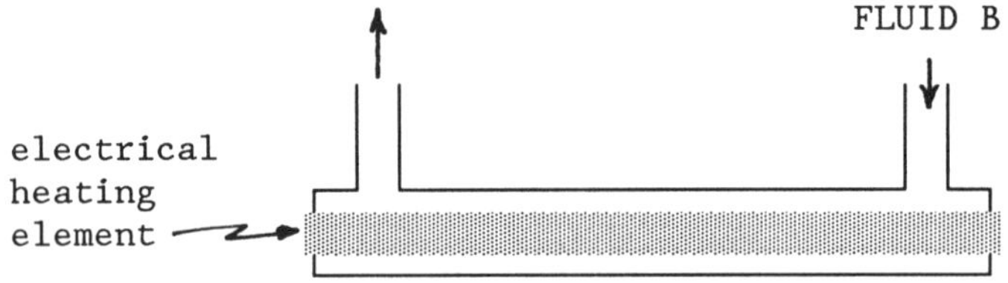

FIGURE (43-5) HEAT EXCHANGER FOR PROBLEM (43-2)

GIVEN

- The electrical power supply for the heating element is a "constant current" type. The value of the current is set by the equipment operator, and is unaffected by the temperature of the heating element. In other words, the current is an independent parameter.

- The power supply is infinitely variable and can be set to supply current in the range 0 to 50 amps.

- The electrical behavior of the heating element is described by Eq. (43-11) where T_h refers to the temperature of the heating element.

$$V = 2.17I(1 - .01(T_h - 100)) \qquad (43\text{-}11)$$

- The temperature drop through the heating element is small and may be neglected.

- The heat flux at the surface of the heating element is given by

$$q = 483VI \qquad (43\text{-}12)$$

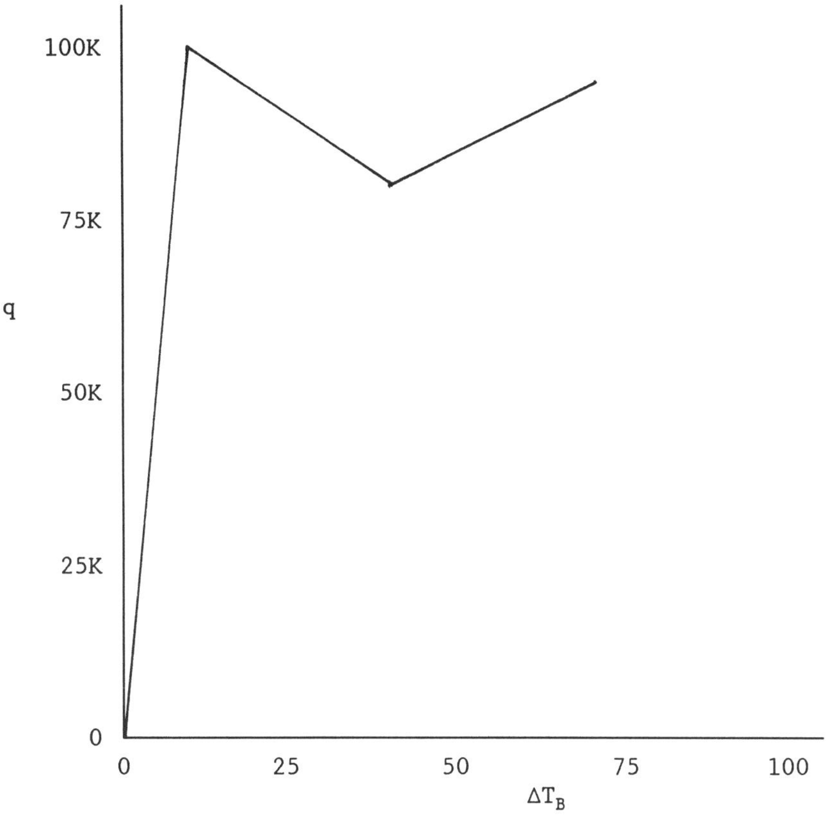

FIGURE (43-6) Thermal Behavior of Fluid B Interface, Problem (43-2)

- The thermal behavior of the Fluid B interface is described by Fig. (43-6). The Fluid B inlet temperature is 100F.
- The temperature rise in Fluid B is small and may be neglected.

ANALYSIS OF PROBLEM (43-2)

Because the temperature drop through the heater is small, and the temperature of Fluid B is 100 F, we may rewrite Eq. (43-11) in the form of Eq. (43-13) where ΔT refers to the temperature difference across the heater/Fluid B interface.

$$V = 2.17I(1 - .01\Delta T) \qquad (43\text{-}13)$$

Since the electrical current is an independent parameter, we may substitute for V in Eq. (43-12) and obtain:

$$q = 1048I^2(1 - .01\Delta T) \qquad (43\text{-}14)$$

(If the power supply had been "constant voltage", we would have substituted for I in Eq. (43-12), and Eq. (43-14) would have been in the form $q\{V,\Delta T\}$.)

Notice that Fig. (43-6) is of the form $q\{\Delta T\}$ and that Eq. (43-14) is of the form $q\{I,\Delta T\}$. Therefore, we can determine $q\{I\}$ by the simultaneous solution of Eq. (43-14) and Fig. (43-6). This is readily accomplished by picking a set of coordinates along the curve in Fig. (43-6), then using Eq. (43-14) to determine the electrical current at each coordinate pair. The q,I coordinates are then used to generate a plot of $q\{I\}$, the desired overall thermal behavior. The details of the analysis are presented in Table (43-2).

TABLE (43-2)

ANALYSIS OF PROBLEM (43-2)

q from Fig (43-6)	ΔT from Fig (43-6)	I from Eq. (43-14)
0	0	0
25000	2.5	4.95
50000	5	7.09
75000	7.5	8.80
100000	10	10.30
90000	25	10.70
80000	40	11.28
85000	55	13.43
90000	70	16.92

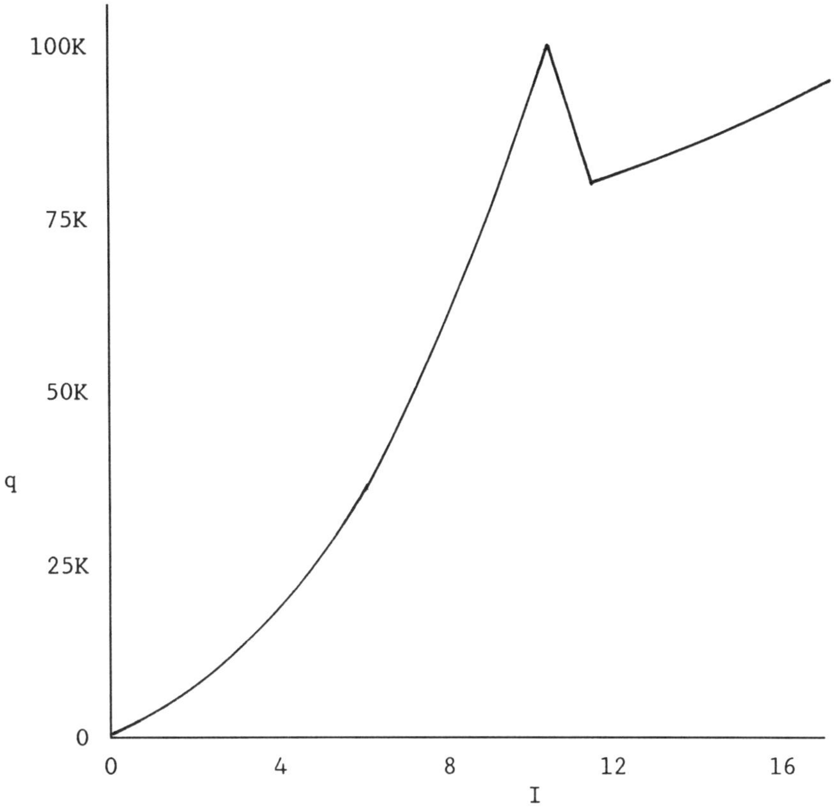

FIGURE (43-7) Overall Thermal Behavior of Heat Exchanger in Problem (43-2)

ANSWER TO PROBLEM (43-2)

Fig. (43-7) is the answer to Problem (43-2). It describes the overall thermal behavior of the subject heat exchanger. The coordinates in Fig. (43-7) are taken from Table (43-2).

DISCUSSION OF PROBLEM (43-2)

The solution of Problem (43-2) would be trivial in the old heat transfer because it overlooks the interaction between the power supply and the temperature of the heated wall. In the old heat transfer, "heat in" curves on $q\{\Delta T\}$ plots are thought to be horizontal if the power supply is electrical, and this is equivalent to stating that the heat input is independent of the temperature of the heated wall.

In the old heat transfer, the solution of Problem (43-2) would require nothing more than the information that the power supply is electrical and that the $q\{\Delta T\}$ behavior of the interface is described by Fig. (43-6). The analysis would involve inspection of Fig. (43-6) to determine the effect of monotonically increasing and decreasing the power to the heater, based on the view that the "heat in" curve is a horizontal line on the figure. The solution would incorrectly indicate temperature discontinuities and hysteresis in the region where $dq/d\Delta T$ is negative.

Problem (43-2) demonstrates how to use the new heat transfer to correctly determine the overall thermal behavior of heat transfer equipment in which the heat source is power controlled. In general, the solution of such problems involves the following steps:

- Determine an expression which describes the relationship among the power input, the independent parameter in the power supply, and the temperature of the heated wall. The desired form is $q\{IP,\Delta T\}$ where IP refers to the independent parameter in the power supply. Eq. (43-14) is an example of this type of expression.

- Obtain the simultaneous solution of the above expression and the figure or expression which describes the thermal behavior of the interface between the heated wall and the fluid. Table (43-2) demonstrates the solution if the thermal behavior of the highly nonlinear interface is described graphically.

- Plot the heat load vs the independent parameter in the power supply. Fig. (43-7) is such a plot.

- Note that temperature discontinuities and hysteresis result only if the heat load is a multi-valued function of the independent parameter in the power supply.

- If the thermal behavior of the heated interface includes a negative slope region bounded by endpoints at which $dq/d\Delta T$ is infinite, and if there is no hysteresis in such a region, indicate that the equipment will behave in an undamped oscillatory manner in this region. Within this region, the heat flux varies between the values at the endpoints of the negative slope region.

In real systems, the truly independent parameter is usually the setting of a knob or a throttle rather than volts or amps, and it is only approximately true that volts or amps is an independent variable. If the power supply cannot be approximately described as either constant voltage or constant amps, the analysis must reflect that the independent parameter is the setting of the knob or throttle. In this case, the desired overall thermal behavior would be a description of how the heat flux or heat load responds to the setting of the knob or throttle.

CONCLUDING REMARKS

The problems in this chapter demonstrate how to use the new heat transfer to determine the behavior of heat transfer equipment which contains a highly nonlinear interface. The first problem demonstrates the analysis when the heat source is temperature controlled (such as a condensing vapor or a flowing fluid); the second problem demonstrates the analysis when the heat source is power controlled (such as electrical or nuclear).

The analysis using the new heat transfer is considerably more difficult than the old way analysis. The old heat transfer analog of this lengthy chapter occupies only a few lines in current texts. The sole reason for the difference in degree of difficulty is that today's old heat transfer analysis is simply incorrect.

The old heat transfer analysis of Problem (43-1) leads to the conclusion that the heat exchanger operates stably throughout its operating envelope because it contains a heat source which is temperature controlled. But this answer is of no value because it is the *wrong* answer.

The new heat transfer analysis of Problem (43-1) leads to the correct conclusion that the heat exchanger operates in an undamped, oscillatory way in one region of its operating envelope, and exhibits pronounced hysteresis in another region.

The old heat transfer analysis of Problem (43-2) leads to the conclusion that the heat exchanger exhibits pronounced hysteresis within its operating envelope because the heat source is power controlled, and because there is a maximum in the $q\{\Delta T\}$ behavior of interface B. But this answer is of no value because it is the *wrong* answer.

The new heat transfer analysis of Problem (43-2) correctly concludes that the heat exchanger operates in a continuous and reversible manner throughout its operating envelope--ie there is *NO* hysteresis, even though the heat exchanger contains a power controlled heat source.

It is certainly true that the new heat transfer analysis of Problems (43-1) and (43-2) is more difficult than the old heat transfer analysis of these same problems. But the new heat transfer analysis results in correct solutions, whereas the old heat transfer results in solutions which are quantitatively and qualitatively incorrect.

In short, the old heat tramsfer simply cannot cope with highly nonlinear heat transfer phenomena. And the new heat transfer can.

CHAPTER 44

THE THERMAL ANALYSIS OF FORCED CONVECTION HEAT TRANSFER EQUIPMENT

In both the old and the new heat transfers, the thermal analysis of forced convection equipment is a two step procedure:

- Generate an expression for the thermal behavior of a differential element--an expression which describes how the local heat flux is related to the local conditions. (In the old heat transfer, generate an expression which describes how the local value of the overall heat transfer coefficient is related to the local conditions.)

- Integrate the expression over the heat transfer surface of the equipment.

Up to this point, the problems have dealt with simple forms of equipment rather than with differential elements. However, the simple equipment forms were selected so that there was no spatial variation in any of the heat transfer parameters. The end result is that the equipment in the preceding problems behaves in precisely the same manner as a differential element, and thus the problems describe how to deal with differential elements as well as with the equipment forms on which the problems are actually based.

Since the problems up to this point have been designed with no spatial variation in the heat transfer parameters, integration has been simply a matter of multiplying the uniform heat flux times the total heat transfer area. Oftentimes, this simple integration is adequate--for example, it is often adequate when dealing with pool boilers.

But when dealing with the behavior of forced convection equipment, integration is seldom a matter of multiplying a uniform heat flux times a total heat transfer area. It generally involves integrating a spatially variable heat flux over the heat transfer surface, and this is a much more difficult problem.

In this chapter, we discuss the analysis of forced convection heat transfer equipment in a qualitative way, with particular emphasis on highly nonlinear behavior. We note that the analysis of this equipment type generally involves integration. Since the integration process is the same in both the old and the new heat transfers, we are not concerned here with integration per se. We are concerned with the impact of the new heat transfer on the overall analysis of forced convection equipment.

THE THREE TYPES OF HEAT TRANSFER INTEGRATION

With regard to integration, heat transfer equipment can be classified in three types:

1. No spatial variation in any parameter which affects heat transfer. In other words, the heat flux is uniform throughout the equipment. (In the language of the old heat transfer, the overall heat transfer coefficient and the overall ΔT are uniform throughout the equipment.)

2. No spatial variation in thermal coupling. (In the language of the old heat transfer, the overall heat transfer coefficient is uniform throughout the equipment.)

3. Spatial variation in thermal coupling. (In the language of the old heat transfer, the overall heat transfer coefficient varies from point to point in the equipment.)

Integration in each of the above groups is accomplished as indicated below:

Type 1: $$\int q \, dA = qA \qquad (44\text{-}1)$$

Type 2: $$\int q \, dA = q_{LM} A \qquad (44\text{-}2)$$

Type 3: $$\int q \, dA = \int q \, dA \qquad (44\text{-}3)$$

Pool boilers are usually Type 1. One phase, forced convection heat exchangers are oftentimes Type 2. Two phase, forced convection heat exchangers are usually Type 3.

Most of the problems in this book demonstrate how to deal with Type 1 equipment using the new heat transfer. However, it is important to note that problems based on Type 1 equipment demonstrate how to deal with a differential element. In this sense, the many Type 1 problems in this book describe the analysis of equipment in a very general way, even though the problems outwardly seem to deal with very simple equipment.

DESIGN CORRELATIONS FOR POOL BOILING AND FORCED CONVECTION BOILING--OLD HEAT TRANSFER

In the old heat transfer, pool boiling data are presented in the form $q\{\Delta T\}$. This form is considered to be an effective way to present pool boiling data, but it is not intended to be a design correlation form. Design correlations are regime oriented, h correlations. In the old heat transfer, there are design correlations for h in the nucleate boiling regime, other design correlations for h in the film boiling regime, and presumably there will one day be design correlations for h in the transition boiling regime.

Cooper (1984) presents a comprehensive summary of design correlations for h in the nucleate boiling regime. He indicates that many of the current correlations are of the form h{q, fluid properties, interface properties}, and that the $h\{q\}$ functionality is described by

$$h \propto q^m \qquad (44\text{-}4)$$

He states: "The average value of m, determined from all data sets, is 0.67." (Note that this is equivalent to stating that q is proportional to ΔT^3, and recall that the nucleate boiling data in the literature indicate q is linearly related to ΔT.)

Forced convection boiling data is seldom (if ever) presented in the form $q\{\Delta T\}$. It is usually presented in the form h vs vapor quality, or in terms of popular dimensionless parameters which separately include h and vapor quality. Design correlations are regime oriented, h correlations in which h may or may not depend on q. The regimes include regimes based on flow pattern--for example, slug flow regime, annular flow regime, mist flow regime. Kenning and Cooper (1989) present a summary of current design correlations used in the old heat transfer.

It is important to note that these boiling h correlations of the old heat transfer generally indicate a monotonic increase in q as ΔT increases--ie they do not address the critically important design problem caused by the maximum in $q\{\Delta T\}$. For this reason, boiling h design correlations cannot be used alone. They must be used in conjunction with other design correlations which describe how q_{max} is related to the parameters in the boiling h correlations. In other words, it is necessary to separately verify that the design q is not greater than q_{max}, since h correlations have nothing to do with the value of q_{max}.

DESIGN CORRELATIONS FOR POOL BOILING--NEW HEAT TRANSFER

In the new heat transfer, $q\{\Delta T\}$ is the design correlation form for all convective heat transfer phenomena, including pool boiling. The pool boiling curve *in its entirety* is the desired correlation. The end result is that:

- We do *NOT* break down the PBC into regimes in order to generate design correlations for each regime.

- We do *NOT* generate design correlations in order to predict $q_{max}\{\Delta T\}$.

- We do *NOT* generate design correlations in order to predict $q_{min}\{\Delta T\}$.

- The PBC design correlation is a complete description of the behavior of the boiling interface because it includes a quantitative description of both the maximum and the minimum in $q\{\Delta T\}$. In other words, no additional correlations are required for critical heat flux or "burnout" because the PBC contains this information.

As the many examples in this book demonstrate, q{ΔT} is the basis for the analysis of heat transfer equipment using the new heat transfer. Pool boiling equipment is no exception, and pool boiling curves are design correlations in the new heat transfer. Pool boiling curves contain all the design information we require, and in a simple form which is easy to understand and to use.

For example, if we are given a particular PBC and asked to design/analyze a pool boiler, we do not need a correlation to predict the value of q_{max} because the PBC explicitly contains this information. We do not need a correlation to predict the value of q_{min} because the PBC explicitly contains this information. And we do not require correlations which describe where the various regimes begin and end because the regimes are implicitly delineated in the PBC, and because we have no need to determine the regime (since we need not decide among several regime oriented design correlations).

DESIGN CORRELATIONS FOR FORCED CONVECTION BOILING--NEW HEAT TRANSFER

In the new heat transfer, design correlations for forced convection boiling are in the generic form q{ΔT}. Forced convection boiling correlations differ from pool boiling correlations mainly in that each correlation is a family of curves rather than a single curve. Each curve in the family describes the thermal behavior at a particular value of vapor quality. The family of curves describes the functionality among q, ΔT, and vapor quality (x, no dimensions) for fixed values of pressure, mass flow rate, the condition of the heat transfer surface, etc.

A pool boiler is very much like a differential element in a forced convection boiler, and we assume that a pool boiling curve closely resembles a forced convection boiling curve for low vapor quality and low mass flow rate.

Adiutori (1965) describes the correlation methodology of the new heat transfer with regard to forced convection boiling:

The (new heat transfer requires) that nonlinear heat transfer phenomena be correlated by determining the functionality between q and ΔT rather than by determining heat transfer coefficients. Let us . . . turn our attention to forced convection boiling which is generally correlated by the use of heat transfer coefficients. We shall attempt to describe the boiling phenomenon by relating q to ΔT graphically and for a single fluid. . . . The problem reduces to establishing the relationship between q, ΔT, and x (vapor quality, no dimensions).

. . . we can at once place two lines on our graph of q vs ΔT:

(i) The line of 100 per cent quality which we know will be highly linear, ie one phase, forced convection.

(ii) The line of 0 per cent quality which we have every reason to expect will be similar to Nukiyama's pool boiling curve.

We therefore begin our graph with the two lines shown in Fig. (44-1). It is of course not possible to fill in the lines of intermediate quality with any great accuracy. However, the 0 per cent line must somehow (and in a continuous manner) assume the shape of the 100 per cent line as the quality increases to that value. Since we cannot at this point demonstrate that the change will be monotonic, let us simply assume that it is, in which case Fig. (44-2) will result.

. . . Fig. (44-2) applies only to a single fluid and was obtained for a particular flow rate and a particular pressure, ie these correlating parameters have been held constant for the sake of simplicity. However, it should be noted that Fig. (44-2) is in precisely the form we require in order to answer the question: How will the evaporator perform at a particular flow rate and pressure for the fluid of interest? Moreover, it may very well be that the effect of these other correlating parameters is small enough to be neglected, in which case Fig. (44-2) becomes more generally applicable.

For instance, the effect of pressure on Fig. (44-2) may be negligible, and Fig. (44-2) might then apply to a wide range of pressures--and similarly with the flow rate. (A footnote states: It has been the author's first-hand experience that boiling liquid metal data are

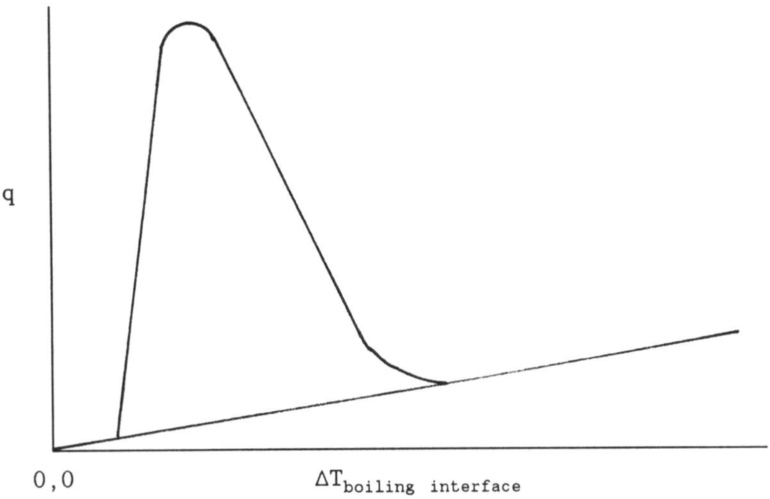

FIGURE (44-1) Forced Convection Boiling--
Thermal Behavior Correlation,
Vapor Quality = 0% and 100%

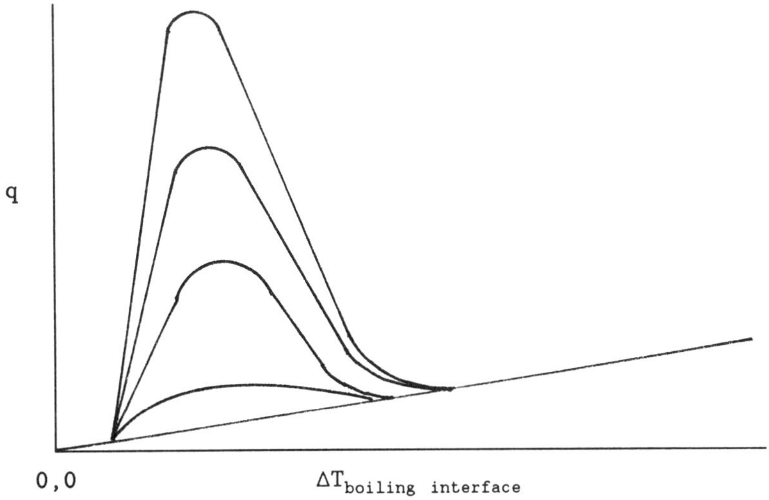

FIGURE (44-2) Forced Convection Boiling--
Thermal Behavior Correlation,
Vapor Quality Parameter

highly correlated in the manner shown in Fig. (44-2) and that, when correlated in this way, the effects of pressure and flow rate are negligible over a considerable range.)

In any event, the correlations resulting from the experimental evidence would be expected to relate q, ΔT, and x in the manner shown in Fig. (44-2) for a given flow rate, pressure, etc. The important point is that Fig. (44-2) correlates the experimental data in the manner required for the effective design of boiling equipment (using the new heat transfer).

REFERENCES

Adiutori, E. F., 1964a, "New Theory of Thermal Stability In Boiling Systems", *Nucleonics*, **22** (5), pp 92-101

Adiutori, E. F., 1964b, "Nucleate Boiling: The Relationship Between Heat Flux and Thermal Driving Force", accepted for publication in the AIChE Journal on April 21, 1964; currently awaiting publication. (An abridged version of this manuscript was published in an advertisement which appeared in Nucleonics, April, 1965.)

Adiutori, E. F., 1964c, "Author's Reply" to "Thermal Stability View Disputed" by numerous disputants, *Nucleonics*, **22** (12), Letters Section

Adiutori, E. F., 1965, "Non-linear Heat Transfer Phenomena", *British Chemical Engineering*, (12), pp 840-845

Adiutori, E. F., 1989, "A New Look at the Origin of the Heat Transfer Coefficient Concept", ASME Paper 89-HT-3, presented at the 26th National Heat Transfer Conference, Philadelphia, PA

Berenson, P. J., 1960, "On Transition Boiling Heat Transfer from a Horizontal Surface", Doctoral Thesis, M.I.T., Cambridge, Mass.

Berenson, P. J., 1962, "Experiments on Pool-Boiling Heat Transfer", Int. Jour. Heat Mass Transfer, **5**, pp 985-999

Bergles, A. E., 1988, "Enhancement of Convective Heat Transfer: Newton's Legacy Pursued", *History of Heat Transfer*, edited by E. T. Layton, Jr. and J. H. Lienhard, ASME Heat Transfer Division

Biot, J.B., 1804, "Memoire sur la propagation de la chaleur", Bibliotheque Brittanaique **37**, 310-329, per Herivel (1975)

Bromley, L. A., 1950, "Heat Transfer in Stable Film Boiling", Chem. Eng. Prog., **46**, pp 221-227, per Kreith and Bohn, (1986)

Braunlich, R. H., 1941, "Heat Transfer to Boiling Liquids Vacuum", Thesis, M.I.T., Cambridge, Mass., per Berenson (1962)

Carne, M. and Charlesworth, D. H., 1965, "Thermal Conduction Effects on the Critical Heat Flux in Pool Boiling", AIChE Preprint 11, Eighth National Heat Transfer Conference, Los Angeles, CA

Cichelli, M. T., and Bonilla, C. F., 1945, "Heat Transfer to Boiling Liquids Under Pressure", Trans AIChE, **41**, pp 755-787

Collier, J. G., 1981, *Convective Boiling and Condensation*, 2nd ed., McGraw-Hill, New York, per Kreith and Bohn (1986)

Colver, C. P., and Balzhiser, R. E., 1965, "A Study of Saturated Pool Boiling Potassium up to Burnout Heat Fluxes", CEP Sym Series, **61**, No. 59

Cooper, M. G., 1984, "Heat Flow Rates in Saturated Nucleate Pool Boiling--A Wide-Ranging Examination Using Reduced Properties", *Advances in Heat Transfer* by Hartnett, J. P., and Irvine, T. F., v 16, pp 157-237

Corty, C., 1951, Ph.D. Thesis, Univ. Mich., Ann Arbor

Corty, C. and Foust, A. S., 1955, "Surface Variables in Nucleate Boiling, CEP Symposium Series, **51**, No. 17, p 1

Drew, T. B., and Mueller, C., 1937, "Boiling", *Trans. AIChE,* **33**, p 449, per Incropera and Dewitt (1985)

Eckert, E. R. G., and Drake, R. M., 1972, *Analysis of Heat and Mass Transfer*, McGraw-Hill, New York

Ellion, M. E., 1954, "A Study of the Mechanism of Boiling Heat Transfer", Memo 20-88, Jet Propulsion Laboratory (Originally prepared as a Ph.D. thesis, Cal. Inst. Tech., Pasadena, CA)

Farber, E.A., and Scorah, R.L., 1948, "Heat Transfer to Water Boiling Under Pressure", Trans ASME, **70**, 369-383

Fourier, J., 1822, *The Analytical Theory of Heat*, English translation by Freeman (1878); republication by Dover, New York (1955)

Halliday, D. and Resnick, R., 1978, *Physics*, 3rd ed., John Wiley & Sons, Inc., New York

Harman, G. G., and Gordy, L. H., 1967, "The Use of Semiconductors for the Study of Boiling Heat Transfer to Low Temperature Liquids", Cryogenics, April, pp 89-92

Hartnett, J. P., Birkebak, R. C., and Eckert, E. R. G., 1961, "Velocity Distributions, Temperature Distributions, Effectiveness and Heat Transfer for Air Injected Through a Tangential Slot into a Turbulent Boundary Layer, ASME Journal of Heat Transfer, August, pp 293-306

Herivel, J., 1975, *Joseph Fourier*, Clarendon Press, Oxford

Hesse, G., 1973, "Heat Transfer in Nucleate Boiling, Maximum Heat Flux and Transition Boiling", Int Jour Heat Mass Transfer, **16**, pp 1611-1627

Holman, J. P., 1986, *Heat Transfer*, 6th ed., McGraw-Hill, New York

Ibele, W., 1963, *Modern Developments in Heat Transfer*, Academic Press, New York

Incropera, F. P., and DeWitt, D. P., 1985, *Fundamentals of Heat and Mass Transfer*, 2nd ed., John Wiley & Sons, New York

Kaulakis, E. G., and Sherman, L. M., 1938, Thesis, M.I.T., Cambridge, Mass., per Berenson (1962)

Kays, W. M., 1966, *Convective Heat and Mass Transfer*, McGraw-Hill

Kenning, D. B. R., and Cooper, M. G., 1989, "Saturated Flow Boiling of Water in Vertical Tubes", Int Jour Heat Mass Transfer, **32** (3), pp 445-458

Klein, J. and Tribus, M., 1953, "Forced Convection from Non-Isothermal Surfaces", Heat Transfer, A Symposium, University of Michigan Press, Ann Arbor, Michigan, per Hartnett, Birkebak, and Eckert (1961)

Kovalev, S. A., 1964, On Stability of Boiling Operating Conditions, *Teplofiz. Vysok. Temp.* **2**(5), 780-788, per Kovalev (1968)

Kovalev, S. A., 1968, "On Methods of Studying Heat Transfer in Transition Boiling", *Int. J. Heat Mass Transfer*, **11** (2), p. 279-283

Kreith, F., 1973, *Principles of Heat Transfer*, 3rd ed, Intext, New York

Kreith, F., and Bohn, M. S., 1986, *Principles of Heat Transfer*, 4th ed., Harper & Row, New york

Levy, S., 1959, "Generalized Correlation of Boiling Heat Transfer", J. Ht Trsfer, **81C**, pp 37-42, per Holman (1986)

Lyon, R. E., Foust, A. S., and Katz, 1955, "Boiling Heat Transfer with Liquid Metals", Chem. Engng. Prog. Symp. Ser. **51**(17), p. 41-47, per Winterton, (1983)

Marto, P. J., and Rohsenow, W. M., 1966, "Nucleate Boiling Instability of Alkali Metals", Trans ASME, **88C**, 183

Maxwell, J. C., 1873, *Electricity and Magnetism*, v. 1, pp 296, 297, per Kraus reprint of 1891 translation of *The Galvanic Circuit Investigated Mathematically* by G. S. Ohm, D. Van Nostrand Co., New York

Maxwell, J. C., 1883, *Theory of Heat*, 7th ed., Appleton, New York

McAdams, W. H., 1948, discussion following presentation of Farber and Scorah (1948)

McAdams, W. H., 1954, *Heat Transmission*, 3rd ed., McGraw-Hill, New York

Mesler, R. B. and Banchero, J. T., 1958, "Effect of Superatmospheric Pressures on Nucleate Boiling of Organic Liquids", AIChE Journal, **4**(1), p 102-113

Newton, I., 1701, "A Scale of the Degrees of Heat", *Phil Trans Royal Soc (London)*, **22**, p 824

Nukiyama, S., 1934, "The Maximum and Minimum Values of the Heat Q Transmitted from Metal to Boiling Water Under Atmospheric Pressure", J. Japan Soc. Mech. Engrs. **37** (207), 367 per Nukiyama 1984 (Also published in English in Int Jour Heat Mass Transfer, v. 9, pp 1419-1433 (1966) and v. 27, No. 7, pp 959-970, (1984))

Nukiyama, S., 1984, "Memories of My Research on Boiling", Int Jour Heat Mass Transfer, v. 27, No. 7, pp 955-957

Ohm, G. S., 1827, *The Galvanic Circuit Investigated Mathematically*, 1969 Kraus reprint of 1891 translation by D. Van Nostrand Co., New York

Papell,S., and Trout, A. M., 1959, "Experimental Investigation of Air Film Cooling Applied to an Adiabatic Wall by Means of an Axially Discharging Slot, NASA TN D-9, per Hartnett, Birkebak, and Eckert (1961)

Perry, C. W., 1940, Ph.D. thesis, John Hopkins Univ., Baltimore, MD, per Mesler and Banchero (1958)

Peterson, W. C. and Zaalouk, M. G., 1971, "Boiling Curve Measurements from a Controlled Heat Transfer Process", ASME Journal of Heat Transfer, pp 408-412

Rohsenow, W. M., 1952, "A Method of Correlating Heat-Trnasfer Data for Surface Boiling of Liquids", Transactions of the ASME, August, pp 969-976

Rohsenow, W. M., 1963, "Heat Transfer with Boiling", *Modern Developments in Heat Transfer*, edited by W. Ibele, Academic Press, New York and London, pp 85-158

Rohsenow, W. M., 1964, Open Letter to E. Adiutori dated April 27, 1964

Rohsenow, W. M., 1973, Boiling section in *Handbook of Heat Transfer* by Rohsenow, W. M. and Hartnett, J. P., McGraw-Hill, New York

Saunders, O. A., 1976, Review of *The New Heat Transfer*, Vols 1 and 2, by Eugene F. Adiutori, *Int. J. Heat Mass Transfer,* **19** (4), p 452

Schnurmann, R. and Lardge, M. G. C., 1973, "Enhanced Heat Flux in Non-uniform Electric Fields", Proc. R. Soc. Lond., **334**, 71

Seban, R. A., 1960, "Heat Transfer and Effectiveness for a Turbulent Boundary Layer with Tangential Fluid Injection, ASME Journal of Heat Transfer, pp 303-312, Nov

Silver, R. S., 1974, Review of *The New Heat Transfer*, 1st ed., Int. Jour. Heat Mass Transfer, Dec, pp 1629-1630

Stephan, K., 1965, "Stabilitat beim Sieden", Brennst.-Warme-Kraft **17** (12), pp 571-578, per Stephan (1968)

Stephan, K., 1966. "Ubertragung hoher Warmestromdichten an siedende Flussigkeiten, *Chemie-Ingr Tech.* **38**(2), 112-117, per Stephan (1968)

Stephan, K. 1968, Letter to the Editors, "On Methods of Studying Heat Transfer in Transition Boiling", Int. J. Heat Mass Transfer, **11**, pp 1735-1736

Stock, B. J., 1960, "Observations on Transition Boiling Heat Transfer Phenomena, ANL-6175

Thome, J. R., and Shock, R. A. W., 1984, "Boiling of Multicomponent Liquid Mixtures", *Advances in Heat Transfer* by Hartnett, J. P. and Irvine, R. F., v 16, pp 59-156

Wallis, G. B., 1965, "A Vote for Adiutori", Nucleonics, **23** (10), pp 6-10

Westwater, J. W., 1959, "Boiling Heat Transfer", Am. Sci., v 47, no. 3, pp 427-446, per Kreith and Bohn (1986)

Westwater, J. W., and Breen, B. P., 1962, "Effect of Diameter of Horizontal Tubes on Film Boiling Heat Transfer", Chem. Eng. Prog., **58**, pp 67-72

Wheatstone, 1843, Bakerian Lecture, per Ohm reprint cited above, p. 51

White, H. E., 1962, *Modern College Physics*, 4th ed., D. Van Nostrand Co., Princeton

Wieghardt, K., 1943, "Ueber das Ausblasen von Warmluft fur Enteisen, ZBW Research Report No. 1900, per Hartnett, Birkebak, and Eckert (1961)

Wieghardt, K., 1946, AAF Translation, No. F-TS-919-RE, August, per Kays (1966)

Winterton, R. H. S., 1983, "Comments on the Existence of Two "Transition" Boiling Curves", Int. J. Heat Mass Transfer, **26** (7), p. 1103-1104

NOMENCLATURE

UNCONVENTIONAL NOTATION

- The symbol "=" means numerically equal. It makes *NO STATEMENT* about dimensional identity.

- Parameter dimensions are implicit in the symbols and are not repeated in the text. For example, if the area is 4.6 ft^2, the text states "A = 4.6", since the nomenclature defines A to mean area *in square feet.*

- y{x} means that y is an unspecified function of x.

- <s means that the system is stable if the inequality is satisfied.

- <u means that the system is unstable if the inequality is satisfied.

SYMBOLS

A	area in ft^2
a	arbitrary constant
a_0	pure constant, numerically equal to thermal conductivity in appropriate units.
b	arbitrary constant
C	electrical capacitance in farads, or heat capacity in Btu/lb F
d	diameter, ft.
D	diameter, ft.
DG	generic dimensionless group
E	modulus of elasticity, psi
f	friction factor
g	gravity constant
h	heat transfer coefficient, Btu/hr ft^2 F
H	height coefficient, ft/lb
I	amps
k	thermal conductivity, Btu/hrftF
L	tank level in inches, or inductance in henries, or height in feet
m	ratio of mass flow rates, slot/mainstream, no dimensions, or arbitrary constant
n	arbitrary constant
Nu	Nusselt number, no dimensions
q	heat flux, Btu/hr ft^2 F

Q	heat load, Btu/hr
P	pressure, psia
PBC	pool boiling curve
Pr	Prandtl number
R	resistance, Ohms
Ra	Rayleigh number, no dimensions
Re	Reynolds number, no dimensions
s	distance, feet, or slot height, inches
St	Stanton number, no dimensions
t	time, secs, or thickness, inches
T	temperature, F
TDF	thermal driving force, F or F/inch
U	overall heat transfer coefficient, Btu/hr ft^2
V	volts or velocity, ft/sec
w	arbitrary variable
W	flow rate, lbs/sec, or weight, lbs
W'	flow rate per slot length, lbs/sec inch
x	distance from slot, inches, or generic distance, inches, or vapor quality, or arbitrary variable
y	arbitrary variable
z	arbitrary variable
ϵ	strain or emissivity, no dimensions
ξ	parameter used in film cooling, no dimensions

SYMBOLS cont

η	parameter used in film cooling, no dimensions
μ	viscosity, lbs/hr ft
ρ	density, lbs/ft^3
σ	stress, lbs/in^2
χ	vapor quality

SUBSCRIPTS

bi	boiling interface
bp	boiler plate
c	coolant
i	ith or inner
LM	log mean
m	mainstream
ms	mainstream
o	outer
s	source
sat	saturated
t	total
w	wall